Tourism Paradoxes

TOURISM AND CULTURAL CHANGE
Series Editors: **Professor Mike Robinson,** *Ironbridge International Institute for Cultural Heritage, University of Birmingham, UK* and **Professor Alison Phipps,** *University of Glasgow, Scotland, UK*

Understanding tourism's relationships with culture(s) and vice versa, is of ever-increasing significance in a globalising world. TCC is a series of books that critically examine the complex and ever-changing relationship between tourism and culture(s). The series focuses on the ways that places, peoples, pasts, and ways of life are increasingly shaped/transformed/created/packaged for touristic purposes. The series examines the ways tourism utilises/makes and re-makes cultural capital in its various guises (visual and performing arts, crafts, festivals, built heritage, cuisine, etc.) and the multifarious political, economic, social and ethical issues that are raised as a consequence. Theoretical explorations, research-informed analyses, and detailed historical reviews from a variety of disciplinary perspectives are invited to consider such relationships.

All books in this series are externally peer-reviewed.

Full details of all the books in this series and of all our other publications can be found on http://www.channelviewpublications.com, or by writing to Channel View Publications, St Nicholas House, 31-34 High Street, Bristol BS1 2AW, UK.

TOURISM AND CULTURAL CHANGE: 57

Tourism Paradoxes

Contradictions, Controversies and Challenges

Edited by
Erdinç Çakmak, Hazel Tucker and Keith Hollinshead

CHANNEL VIEW PUBLICATIONS
Bristol • Blue Ridge Summit

DOI https://doi.org/10.21832/CAKMAK8120
Library of Congress Cataloging in Publication Data
A catalog record for this book is available from the Library of Congress.
Names: Çakmak, Erdinç, editor. | Tucker, Hazel – editor. |
  Hollinshead, Keith, editor.
Title: Tourism Paradoxes: Contradictions, Controversies and Challenges/
  Edited by Erdinç Çakmak, Hazel Tucker and Keith Hollinshead.
Description: Blue Ridge Summit: Channel View Publications, 2021. | Series:
  Tourism and Cultural Change: 57 | Includes bibliographical references
  and index. | Summary: 'At a time when COVID-19 is transforming the
  tourism industry, this book presents many contemporary inconsistencies
  and paradoxes in tourism contexts and studies. It offers a reconsideration
  of what may be needed in order to equip researchers and practitioners in
  tourism and related fields to better interpret and manage the
  future of tourism' – Provided by publisher.
Identifiers: LCCN 2020036900 (print) | LCCN 2020036901 (ebook) |
  ISBN 9781845418113 (Paperback) | ISBN 9781845418120 (Hardback) |
  ISBN 9781845418137 (PDF) | ISBN 9781845418144 (ePub) |
  ISBN 9781845418151 (Kindle Edition)
Subjects: LCSH: Tourism.
Classification: LCC G155.A1 T5924315 2021 (print) | LCC G155.A1 (ebook) |
  DDC 306.4/819 – dc23
LC record available at https://lccn.loc.gov/2020036900
LC ebook record available at https://lccn.loc.gov/2020036901

British Library Cataloguing in Publication Data
A catalogue entry for this book is available from the British Library.

ISBN-13: 978-1-84541-812-0 (hbk)
ISBN-13: 978-1-84541-811-3 (pbk)

Channel View Publications
UK: St Nicholas House, 31-34 High Street, Bristol BS1 2AW, UK.
USA: NBN, Blue Ridge Summit, PA, USA.

Website: www.channelviewpublications.com
Twitter: Channel_View
Facebook: https://www.facebook.com/channelviewpublications
Blog: www.channelviewpublications.wordpress.com

Copyright © 2021 Erdinç Çakmak, Hazel Tucker, Keith Hollinshead and the authors of individual chapters.

All rights reserved. No part of this work may be reproduced in any form or by any means without permission in writing from the publisher.

The policy of Multilingual Matters/Channel View Publications is to use papers that are natural, renewable and recyclable products, made from wood grown in sustainable forests. In the manufacturing process of our books, and to further support our policy, preference is given to printers that have FSC and PEFC Chain of Custody certification. The FSC and/or PEFC logos will appear on those books where full certification has been granted to the printer concerned.

Typeset by Riverside Publishing Solutions.
Printed and bound in the UK by the CPI Books Group Ltd
Printed and bound in the US by NBN.

Contents

Figures and Tables   vii
Contributors   ix
Acknowledgements   xiii
Foreword by Erik Cohen   xv

1 Introduction: Tourism Paradoxes – Contradictions, Controversies and Challenges   1
Erdinç Çakmak, Hazel Tucker and Keith Hollinshead

2 The Paradox of Modernity: Power, Identity and Tourism in Rural Cyprus   15
Evi Eftychiou

3 Go West! Overcoming the Paradoxes of Kinh Tourism in the Vietnamese Mountains: A Postcolonial Geography   33
Emmanuelle Peyvel

4 The 'Logical Paradox' of Preservation via Change: The Touristic Potential of Malaysia's Catholic Mission Schools   50
Keith Kay Hin Tan and Paolo Mura

5 Empowering Package Tour Travellers by Disempowering Tourism Operators? Assessing the Effectiveness of the Tourism Law of China   74
Nan Chen, Kevin Burns and Jing Wang

6 Cross-cultural Encounter: Sustaining Racial Prejudice or Prompting Reflection?   97
Man Tat Cheng

7 Contemporary Polemics of Chinese Outbound Tourism to Europe: Paradoxes, Inconsistencies and Contradictions   114
Rose de Vrieze-McBean

8 International Tourism Academia: A Paradoxical
   Challenge   128
   Vincent Platenkamp

9 The Call for 'Dynamic Genesis' (after Deleuze) in
   Tourism Studies   145
   *Keith Hollinshead, Rukeya Suleman, Sisi Wang,
   Bipithalal Balakrishnan Nair and Alfred Bigboy Vellah*

10 Afterword: Reflections on Paradoxes in Understanding,
   Culture, Mobility, and Tourism   159
   Erdinç Çakmak, Keith Hollinshead and Hazel Tucker

   Index   165

Figures and Tables

Figures

2.1	Old and new Kakopetria village	25
2.2	A building in old Kakopetria, showing the alterations that owners have made	27
3.1	While mountains cover two-thirds of the country, Kinh tourism only occupies a handful of sites: Lunar New Year gathering in the Perfume Pagoda	38
3.2	Gardens, the orderly nature of Kinh mountain tourism	39
3.3	Postcolonial primitivism in Bà Nà	44
3.4	The colony as a romantic setting in Bà Nà	46
4.1	Ruins of St Paul's Church, Malacca, earlier part of a Jesuit school started in 1548	52
4.2	Fairclough's 'Three-dimensional CDA model' (as interpreted by main author)	58
4.3	A 'Three-dimensional CDA model' for narrative interviews	66
4.4	Main block of the Cameron Highlands Convent, built in 1935 on leasehold land, where the district council encourages the use of mock-colonial architecture	69
4.5	Former St Joseph's Novitiate in Penang	70
5.1	Business models for normal GPTs and LFPTs (adapted from Jia, 2006)	77
6.1	Chairman Mao is the great liberator of the world's revolutionary people	107
6.2	The feelings of friendship between the peoples of China and Africa are deep	108
8.1	Three modes of knowledge production	138

Tables

4.1	Research participants	55
5.1	Interviewee profiles	80

Contributors

Kevin Burns PhD is a Lecturer in Tourism Management, Department of Hospitality Management, Dundalk University of Technology, Ireland, having previously lectured in mainland China and Hong Kong. A doctoral graduate from Hong Kong Polytechnic University, his research focuses on tourist behaviour with an emphasis on cultural dimensions, misbehaviours, and destination management. He has co-written four book chapters on cultural tourism and has recently co-published in the *Journal of Travel Research*. He actively presents his research at international and national conferences.

Erdinç Çakmak PhD is Senior Lecturer, Academy for Tourism, Breda University of Applied Sciences, the Netherlands. He is (co-)editor of a number of special volumes and books and has (co-)authored more than 30 academic papers and book chapters. He has (co)-chaired international conferences and special sessions on the topics of power relations in tourism, conflict-ridden destinations, and tourism paradoxes. He has been vice president of the international tourism group RC50 at the International Sociology Association since 2014.

Nan Chen PhD is Research Fellow, School of Hotel and Tourism Management, Hong Kong Polytechnic University. She obtained her PhD from Griffith University, Australia. She has worked as an Assistant Professor at Surrey International Institute and also as a Senior Lecturer at University of Huddersfield, UK. Her research interests include tourism and event management, leisure consumer behaviour, resident-tourist interaction, and emotional experience. Nan's research has been published in leading academic journals, including *Tourism Management, Journal of Travel Research, Journal of Hospitality and Tourism Research* and the *International Journal of Contemporary Hospitality Management*.

Man Tat Cheng PhD began his academic career at Macau University of Science and Technology, China. He is interested in cultural authority and individuals' behaviour patterns in cross-cultural interaction. His recent work focuses on political economy analysis of cross-strait tourism activities. Concurrently, he is developing ideas on the shaping of Chinese cultural identity, first explored in his doctoral thesis.

Evi Eftychiou PhD is Lecturer in Social Anthropology, Department of Law, University of Nicosia, Cyprus. Her publications include articles, book chapters and technical reports in the domains of tourism and environment. During her career, Evi has served as a Research Fellow and as Head of the Erasmus Office, University of Nicosia. She has wide experience of research projects from the more than 35 projects she has coordinated and/or participated in, funded by national, international, and European agencies. Her research interests revolve around identity politics, environment, tourism, education, and human rights.

Keith Hollinshead PhD is a specialist in the politics of culture, be it in 'Indigenous culture' or in or the contemporary representation of longtime traditions/emergent transitions via 'cultural tourism'. He inspects the cherished inheritances which are inscribed through public culture and via worldmaking vistas of place and space; he works on adisciplinary/postdisciplinary fronts via political science, cultural studies, and human communication understandings, inspecting the contested ways in which pasts/places are naturalized and fantasmatic aspirations normalized. Anglo-Australian, he draws upon management experience in Wales, Australia, the USA, and beyond, to critically interpret the declarative activity of 'state-' or 'population-'making via thoughtlines emanating from (mainly) Foucault/Deleuze/Bhabha/Kirschenblatt-Gimblett/Braidotti. A Distinguished Professor of the International Tourism Studies Association (based at Peking University), he is now an Independent Scholar based in both Cheshire and Warwickshire in the UK. He is a Masthead Editor for both *Tourism Analysis* and *Tourism, Culture and Communication*.

Rose de Vrieze-McBean is a researcher, educationalist, linguist, and is Lecturer of English and Academic Skills, Breda University of Applied Sciences, the Netherlands. She holds a PhD in Tourism Management from the University of Bedfordshire (UK). She has multi-faceted areas of interest, which include Chinese outbound tourism, course-design, cross-cultural and social studies and English language acquisition, particularly in young adults. Rose's current research is examining the Next Tourism Generation, a collaborative European research consortium, of which Breda University (of Applied Science) is a member.

Paolo Mura holds a PhD in Tourism from the University of Otago, New Zealand. An Italian by passport, he has lived and conducted research in Germany, Greece, the USA, New Zealand, Malaysia and the United Arab Emirates. He is currently Associate Professor in Tourism, College of Communication and Media Sciences, Zayed University, Abu Dhabi,

UAE. His research interest is tourist behaviour, with a focus on young tourists' experiences, gender, travelling subcultures, and qualitative approaches to research.

Bipithalal Balakrishnan Nair PhD is a transdisciplinary researcher in Tourism Studies with particular interests in postcolonialism and colonial nostalgia, in worldmaking normalizations, and in creative management/ development practices in the representation of peoples and places. Her doctorate in Tourism Management, obtained at the Institute of Tourism Research (University of Bedfordshire, UK), examined soft power orientations to travel and heritage in her native India. Dr Nair has recently joined the University of Woosong in South Korea, where she is Assistant Professor in Tourism and Hospitality Management.

Emmanuelle Peyvel PhD is Associate Professor of Geography, University of West Brittany (Department of Tourism Studies), and a member of the Institute of East Asian Studies (France). Since 2005 her research has explored the development of tourism and leisure in Vietnam, especially in their relation to globalization and (post)socialist nation building. Her work uses a postcolonial and intersectional approach to study unequal access to tourism resources. She regularly teaches tourism studies at Van Lang University and the National University of Economics (Ho Chi Minh City).

Vincent Platenkamp PhD is Emeritus Professor in Cross-Cultural Studies, Academy for Tourism, Breda University of Applied Sciences, the Netherlands. He has been involved in developing international educational and training curricula and teaching on tourism and leisure in more than ten countries in Europe, South America and Asia. Vincent has published broadly in tourism journals on contexts in tourism, mode 3 knowledge production, tourism education, cross-cultural issues, and philosophy.

Rukeya Suleman is a PhD candidate at the University of Bedfordshire, UK, and is currently completing her doctoral work on Islamic identities vis-à-vis globalization. She is a cultural geographer specializing in the interfaces between spirituality, travel, and modernity. Schooled at Cambridge University, Rukeya is a creative thinker on emergent critico-interpretive/ soft-science outlooks in both the humanities and the posthumanities, especially with regard to traditional/transitional constructions of place/space. Principally partnering Keith Hollinshead on geopolitical research agendas, she has published on public culture, worldmaking, and indigenous issues. She was elected joint Vice-President (International Tourism) of the International Sociological Association in 2018.

Keith Kay Hin Tan PhD is a UK-registered architect and co-editor of *Contemporary Asian Artistic Expressions and Tourism* (2020). He obtained his doctorate in Tourism Studies from Taylor's University in Malaysia in 2017, where he is Senior Lecturer, School of Architecture, Building and Design. He has published research articles connecting architecture, tourism and heritage in a variety of international journals and is also an avid journal reviewer.

Hazel Tucker PhD is Professor of Tourism, University of Otago, New Zealand. She holds a doctorate in social anthropology from Durham University, UK, and is the author of *Living with Tourism* (2003) and co-editor of *Tourism and Postcolonialism* (2004) and *Commercial Homes in Tourism* (2009). She is Co-president of the International Tourism Research Committee RC50 of the International Sociological Association and serves as an Associate Editor for *Annals of Tourism Research*. Her research interests include tourism encounters, emotion, heritage, and gender.

Alfred Bigboy Vellah is a PhD candidate at the University of Bedfordshire (UK), where he is completing his doctoral work on the worldmaking issues involved in the projection of Africaneity. He is a human communications commentator who inspects the ways in which different populations/institutions/interest groups imagine the world. Specializing in acts of othering and ethnocentrism, Alfred analyses how 'peoples' and 'places' are composed within longstanding and emergent 'storylines'. Native to Zimbabwe, he works with a political regard over matters of agency and authority involved in (particularly) the representation/misrepresentation of African inheritances.

Jing Wang PhD is Lecturer in Law and a faculty member, School of Law, University of Strathclyde, Glasgow, UK. As a researcher at Strathclyde's Centre for Antitrust Law & Empirical Study (SCALES), Dr Wang conducts research at the broad interface of commercial law/competition law. With a particular interest in consumer welfare protection, Dr Wang's research focuses on the interrelated topics of market intervention, market competition and consumer interests; tourism law and consumer protection; and competition regulation and enforcement in China.

Sisi Wang is a PhD candidate at the University of Bedfordshire (UK) studying contemporary representations of China within international tourism and related inscriptive industries. Currently investigating found differences between so-called 'Eastern' and 'Western' understandings of the world, she pries into 'soft power' articulations of and for 'China'. Sisi's own ongoing research interests include Confucianism/neo-Confucianism/cultural diplomacy, each positioned within the context of tourism and its cousin projective/collaborative industries. She is an active member of the Public Culture (Studies) Group at the University of Bedfordshire.

Acknowledgements

We would like to thank everybody who partook in bringing this volume together. First, we would like to acknowledge the important part played by the International Sociological Association (ISA) RC50, the 'International Tourism' Research Committee which brought the editors together to work on this volume. The idea and planning for the book was initiated while we were in Chiang Mai, Thailand, for the 2016 RC50 'In-between' Conference, and so we would like to thank all those involved in that conference. Next, we are greatly indebted to the contributors, who have produced thought-provoking and challenging ideas on the paradoxes of tourism. We very much appreciate their efforts, time, assistance, under-standing and, at times, their patience, with our requests for details and adjustments. The breadth of viewpoints, together with their stimulating and detailed knowledge of their very different subject matters, have provided unique and wide-ranging topics related to the theme of this volume. We are also grateful to all at Channel View for their support and cooperation throughout the preparation of the book and the submission of the manuscript, and in particular Sarah Williams for her continued encouragement, support and patience. Last but not least, we thank all friends, family and colleagues, in particular RC50 Vice-President Rami K. Isaac, who provided useful comments on the structure of this volume along with much needed encouragement, and Hazel's son, Liam, for producing the Penrose triangle image for the book's cover.

Breda (The Netherlands), Dunedin (New Zealand)
and Nuneaton (England)
November 2020

Foreword

A paradox is a statement which seems self-contradictory but expresses a possible truth (Dann, 2017). Though modern tourism is riven by paradoxes, the late Graham Dann was the first to publish a conceptual paper on paradoxes and oxymora in tourism studies; but he paid more attention to the latter than to the former.

In order to draw attention to this little studied topic, the present volume brings together a collection of chapters on its various aspects. While some of the chapters focus on the deeper philosophical and theoretical facets of the issue, others deal with concrete examples of paradoxes in tourism, revealed in the authors' empirical studies.

Tourism Studies researchers have engaged in the study of paradoxes for a long time, even if they did not explicitly point them out. It is indeed remarkable that the issue of paradoxes in tourism only came up some forty years after the publication of one of the foundational papers in the sociological study of tourism, Dean MacCannell's highly influential 'Staged authenticity' (1973), which implicitly revealed a fundamental paradox inherent in modern tourism. MacCannell's principal thesis was that modern tourists strive for authentic experiences on their trip, but their very desire induces the locals to present them with a staged authentic front, while hiding their real life in an un-accessible back. But though the paradox inherent in this process was there for all to see, nobody to the best of my knowledge made it explicit.

MacCannell's seminal idea has found resonance in the psychological studies of Larsen and his collaborators (e.g. Doran *et al.*, 2015), who found that, paradoxically, tourists think of themselves as being authentic, or individualistic travelers, and dissociate themselves from typical mass tourists, even though they themselves engage in mass tourism.

According to this line of thought, modern (Western) tourism is thus doubly paradoxical: it leads to the staging of the very sites and sights, treasured for their alleged authenticity, even as mass tourists insist on the uniqueness of their individual experiences, though in fact all share the same staged attractions. However, the researchers did not explicitly dwell upon the paradoxical nature of their findings.

This volume is the first to take up the topic of paradoxes in tourism in a focused way, and includes empirical case studies of specific paradoxes inherent in concrete touristic situations as well as two chapters that explore the deeper philosophical aspects of paradoxes in tourism. The editors should be lauded for their efforts to bring together a collection of writings in this book that will hopefully help to put the study of tourism paradoxes on the agenda of tourist studies.

Erik Cohen
Emeritus Professor, Department of Sociology
and Anthropology, Hebrew University of
Jerusalem, Israel

References

Dann, G.M.S. (2017) Unearthing the paradoxes and oxymora in tourism. *Tourism Recreation Research* 42 (1), 2–10.
Doran, R., Larsen, S. and Wolff, K. (2015) Different but similar: Social comparison of travel motives among tourists. *International Journal of Tourism Research* 17, 555–563.
MacCannell, D. (1973) Staged authenticity: Arrangements of social space in tourist settings. *American Journal of Sociology* 79 (3), 589–603.

1 Introduction: Tourism Paradoxes – Contradictions, Controversies and Challenges

Erdinç Çakmak, Hazel Tucker and Keith Hollinshead

> Paradox: But one must not think ill of the paradox, for the paradox is the passion of thought, and the thinker without paradox is like the lover without passion: a mediocre fellow
>
> Søren Kierkegaard (*Philosophical Fragment*)

The idea for this book first came about when the editors met in Chiang Mai, Thailand, for a conference on tourism paradoxes. The location was fitting due to Chiang Mai – which literally means *new city* despite it being established more than seven centuries ago – being full of tourism paradoxes. Now attracting both international tourists and migrants, the provincial capital city of Northern Thailand became a popular center for international backpacker tourists during the 1970s. These 'alternative' youth tourists, as Erik Cohen (1989) called them, were said to be seeking adventurous and unique experiences of 'authentic' primitive life in the hill tribe villages surrounding Chiang Mai. In the words of Dean MacCannell (1989: 2), this '"modern" thirst for authenticity met by a "primitive" capacity to produce dramatic representations of pseudo authenticity' gave rise to a new tourism-related set of living arrangements which was simultaneously, and we could add paradoxically, post-traditional and post-modern. Since then, with a tourism sector that has relatively low entry barriers and offers low/semi-skilled jobs, Chiang Mai has become the second largest city in the country, with migrants coming from surrounding rural areas as well as neighboring countries such as Myanmar and Laos. This tourism and associated economic growth has brought its challenges, however, since the formal economy and infrastructure in Chiang Mai lack the necessary capacity to absorb these newcomers, and a significant informal tourism sector has hence developed. These developments and challenges have led in turn, as is the case in many other tourism destinations around the world, to many new forms of contradictory and paradoxical tourism practices and effects.

In relation to the ever-changing tourist markets, a significant trend in recent years has been a rapid increase in Chinese tourists visiting Chiang Mai following the popularity of a Chinese 2012 road movie, *Lost in Thailand* (see Budde *et al.*, 2013), which was filmed in Chiang Mai. This new influx has brought yet further unexpected challenges and conflicts, both within Chiang Mai's tourism and hospitality sector and in relation to the broader resident 'host population' of the city. In 2015, a discussion forum and exhibition entitled 'My Chiang Mai' was organized by the Thailand Creative and Design Centre (TCDC), aimed at helping the city's residents 'adapt and cope' as they found themselves increasingly 'playing host to this new and unfamiliar group of visitors' (TCDC, 2015). The exhibition presented the question: 'Where can a balance be found between the conflicting interests of those who live here, those who do business with tourists and those who come here as visitors? Can Chiang Mai be both a great place to live and a great place to visit?' (TCDC, 2015). The exhibition had displays intended to help Chiang Mai residents better understand this new tourist group, which rapidly became the main tourist 'market' visiting the city. Recognizing that many around the world are 'hankering' to attract Chinese tourists, the exhibition also alluded to a rapid rise in tensions between Chiang Mai residents and Chinese tourists, and especially to 'locals' complaints about Chinese tourists' lack of manners' (TCDC, 2015). Interestingly, the exhibition's displays were written in both Thai and English, raising the question as to whether it was expected that the exhibition might also be visited by Western tourists, and perhaps the ever-increasing number of Western residents, in order to also address tensions and lack of understanding between Chiang Mai's 'traditional' visitor group and its new visitors. It seems paradoxical that Thai residents and Western tourists in Chiang Mai have become more attuned to each other than either party feels in its relation to the 'new' Chinese tourists. However, if a paradox is something 'seemingly absurd or contradictory which when investigated may prove to be well founded or true' (Soanes & Stevenson, 2006), then many of the Chiang Mai tourism happenings in recent years appear paradoxical indeed. Chiang Mai thus appropriately illustrates the main theme of this book.

While, to date, the field of tourism studies has covered much disciplinary, thematic and methodological ground and forged significant conceptual and practical benchmarks, tourism at both the local and global level was said recently to have entered something of a 'new era', in relation to changing tourist markets as well as to global digital technology and social media trends. It was especially the already alluded to rapid rise in the Asian tourism markets that Winter (2009) referred to as signifying this 'new era' and which placed tourism scholarship at a philosophical and ethical crossroads. Indeed, the global tourism landscape has changed significantly during the last two decades, with Asia shifting from being predominantly a destination for Western tourists to becoming a key tourist-generating region. According to the UNWTO, in 2017 one out of four tourist trips originated in Asia and the Pacific (UNWTO, 2018), with China ranked number one in the world in terms of international tourism expenditure ($US257.7 billion). Also, as already alluded to, this changing landscape gave rise to new tensions and controversies, increasing use of terms such as 'overtourism' and seemingly paradoxical ideas around 'demarketing' and 'degrowth'.

Now, at the time of putting the final touches to this volume, the world is facing yet another new era, this time as a result of the COVID-19 pandemic. Suddenly, 'overtourism' has become 'undertourism', and a great many countries around the world are plunged into thinking about recovery and resilience. This book's focus on tourism's paradoxical complexities thus seems even more topical. Thankfully, while the tensions, contradictions and controversies inherent in tourism only seem to increase in prevalence, there has simultaneously been an increased maturity and sophistication in many realms of tourism production and consumption, as well as an increased reflective awareness of the representational 'powers' of, and ability to deploy, tourism to make, de-make and remake places and peoples. By identifying these kinds of incongruent and paradoxical tourism contexts, as well as approaches in tourism studies, this book thus aims to prompt a reconsideration of what may be needed, conceptually and methodologically, in order to equip tourism studies and related social science fields to work with, and to interpret, tourism's ongoing dynamics. Indeed, as quoted in the philosophical fragment above, one must not think ill of the paradox, for the paradox is the passion of thought.

Centering on the notion of tourism paradoxes has thus invited the authors of this book's chapters to highlight and relish the contradictions and inconsistencies apparent in tourism contexts and tourism studies. While the chapters were all written prior to the COVID-19 pandemic, the book's chapters emphasize the ever-dynamic, ever-present and open nature of tourism's inconsistencies. Overall, the book thus reflects, and aims to further, the growing understanding of a need for tourism studies to focus more on these messier and inconsistent matters of 'becoming' in tourism (see Chapter 9 for discussion on 'becoming'). Other recent works that have similarly embraced the messy and paradoxical aspects of tourism include: *Travels in Paradox, Tourism Encounters and Controversies, Disruptive Tourism and its Untidy Guests,* and *The Practice of Sustainable Tourism: Resolving the Paradox.* Some of these, particularly the latter in this list (Hughes *et al.*, 2015), express a sense of trouble and challenge from tourism's 'seemingly absurd or contradictory' elements (Soanes & Stevenson, 2006), and therefore set out to try to find resolutions to the paradoxes that are especially inherent to, and pervasive within, concepts such as 'sustainable tourism'. *Tourism Encounters and Controversies: Ontological Politics of Tourism Development* (Jóhannesson *et al.*, 2016), in contrast, examines the material and social ontological politics of tourism development in order to *embrace* the ways in which tourism is always relational, unstable and messy. Other books are more 'playful' with the paradoxical ideas inherent in tourism. For example, *Travels in Paradox* (Minca & Oakes, 2006) plays with the dynamic and fluid nature of places to argue that the paradoxes of travel are also the paradoxes of place, including the place of 'home'. *Disruptive Tourism and its Untidy Guests* (Veijola *et al.*, 2014) is also somewhat playful in suggesting that as well as catering to the paying and orderly client, tourism must also welcome its inherent disorderly, disruptive and we could say paradoxical, elements.

The present volume similarly embraces tourism's untidy paradoxes and, embedded within sociological and anthropological scholarship, focuses on some tourism paradoxes, controversies and inconsistencies that have dynamic *cultural* relationships at their center. The chapters are aimed at highlighting the point that many aspects of the relationship between tourism and culture, tourism and social change, tourism and globalization/ glocalization, and tourism and (post)modernity, encompass complex contradictory, and often incongruent, approaches and processes. The overall purpose of the book is therefore to enhance scholarship in this field by encouraging those studying and researching tourism studies to critically engage with the leading salient complexities of tourism and thereby to further embrace and work both with and within some key paradoxical themes, or areas, in relation to tourism and tourism studies. These paradoxical areas include: North Atlantic centrism vis-à-vis non-Western imperatives; established political apparatus vis-à-vis peoples' empowerment; continued colonization vis-à-vis post-colonization and decolonization; fixed/singular identities vis-à-vis liquid/plural aspirations; and globally standardized vis-à-vis locally dynamicized. These are key themes, which are identified and drawn upon throughout the remainder of the book and they will now, each in turn, be outlined in more detail.

Anglo-Western Centrism vis-à-vis non-Western Imperatives

This first paradoxical theme refers to the point that, despite significant changes in the global tourism landscape as mentioned above, tourism theory continues to remain largely rooted in the 'Western tourist gaze' (Chang, 2015; Cohen & Cohen, 2015; Tucker & Zhang, 2016; Winter, 2009; Zhang, 2018). With Asia having become a major, if not *the* major, tourist-generating region, and China now ranking first in the world in terms of international tourism expenditure, many scholars are now wondering whether the Anglo-Western centrism in tourism knowledge means that the field is, in fact, ill-equipped to respond to increased Asian mobilities. Anglo-Western centrism in tourism studies is prompted both in the fact that the 'tourist' that research has mainly focused on is still very often conceived as coming from the Western, industrialized countries (Huang *et al.*, 2014; Keen & Tucker, 2012; Winter, 2009), and in the fact that the producers of tourism knowledge are largely Western scholars. The paradox is, consequently, that despite the changing global tourism landscape, we are still seeing the uncritical application of Anglo-Western theory to explain non-Western tourism phenomena, which in turn leads to 'misguided claims of universality' (Winter, 2009: 23). Hence, there are increasing calls to 're-center', or in other words to do something to 'Asianize' and indigenize, the tourism field (see, for example, Chambers & Buzinde, 2015; Chang, 2015). These calls look to the incorporation of 'non-Western' knowledge and theory as the answer, arguing, as Pritchard and Morgan (2007) put it, that 'we must act to decentre the tourism academy and respond to the challenges and critiques being articulated by indigenous scholars so that we may begin to create knowledge centred on indigenous epistemologies and ontologies' (Pritchard & Morgan, 2007: 22).

However, such attempts to 're-centre' and to indigenize or 'Asianize' the field may become paradoxical in themselves, in that they can often function to further entrench the prevalent binaries and ways of thinking which separate 'West' and 'non-West', even though the above-mentioned 'new era' of tourism itself serves to subvert this binary. In response to this paradox, Cohen and Cohen (2015) suggest using a mobilities paradigm to study the emerging markets in order to be more 'attentive to differences within and between countries in the emerging regions, as well as to similarities between some of these and Western ones' (Cohen & Cohen, 2015: 2). Indeed, Zhang (2018) similarly addresses this paradox in her asking: 'are we really so different from one another? Are Chinese tourists really so different from other tourists?' (Zhang, 2018: 132). Tucker and Zhang (2016) also problematize the idea of encouraging 'alternative discourses', suggesting that such practices run the risk of further entrenching dualisms which, ultimately, may hinder the ability of the tourism field 'to proceed to more open and pluralistic dialogues' (Tucker & Zhang, 2016: 252). These are paradoxical matters indeed and they are picked up again in many of the chapters in this book.

Continued Colonization vis-à-vis post-Colonization and Decolonization

Related to the above theme, the next paradoxical theme worth introducing here is that of tourism's inherent continuing colonization vis-à-vis increasing calls for decolonization of the field as well as attempts to highlight what might be considered 'post-colonial' subversions of tourism's colonizing tendencies. Indeed, a key paradox here is that contemporary tourism practice and much tourism scholarship echo and thus perpetuate colonial discourse, whilst at the same time, 'post-colonial agents' may use tourism in order to counter the narrative accounts of 'colonizing agents'. The notion of global tourism as neo-colonialism has been widely discussed, for example, by Edensor (1998), Jaakson (2004) and Tucker and Akama (2009), and it is argued that in the context of contemporary international tourism, *and* tourism scholarship itself, the economic structures, cultural representations and exploitative relationships that were previously based in colonialism are far from over (Kothari, 2015). Hence, the recent calls, as referred to above, to *de*-colonize the tourism field (Chambers & Buzinde, 2015).

Concurrently, however, it is important to recognize colonialism's own areas of ambivalence and internal contradiction, as well as to consider the role that tourism can play in contesting and reconstructing the legacies of colonialism. The paradox is that, whilst 'culture' has often become essentialized for tourism in accordance with colonial narratives, tourism may at times be used to counter, or subvert, the colonial narrative (Hollinshead, 2004). Hence, a state of paradoxical ambivalence often prevails in relation to how the cultural legacies of colonialism play out in tourism, and just as it is crucial to recognize the continued colonization, it is equally important to recognize localized forms of challenge and resistance manifested through tourism's many subversive cultural assertions (Park, 2016). Indeed, tourism, and 'heritage tourism' in particular, has often been used by post-independence states to assert a new decolonized identity by contesting and readdressing situations of social, cultural and political domination that had arisen through colonialism. Paradoxically, governments of post-independence states may even use heritage tourism to appropriate the language of the colonizer in order to 'write back' (Ashcroft *et al.*, 1989), and thereby 'to respond to and "de-scribe" the discourses of the coloniser' (Marschall, 2004: 102). The matters of continued colonization vis-à-vis post-colonization and decolonization in tourism are referenced in multiple chapters of this book, particularly Chapters 3, 8 and 9.

Established Political Apparatus vis-à-vis Peoples' Empowerment

The third paradox in tourism studies addressed in this volume is that the political apparatus does not always run parallel to the civil apparatus (Gramsci, 1971). While for Foucault (1995), 'power is everywhere' and power relations are embedded in the social structure, Gramsci argues that power is operated through the mutual interactions of economy, culture and politics within the realm of a hegemonic discourse (Daldal, 2014; Jones, 2006). In Gramsci's term, power is mainly exerted by the dominant groups, by working on the popular mentality via the institutions of civil society and establishing a hegemony using the political apparatus. However, this political apparatus is not always fortified by a civil consensus on the level of cultural power that consists of people's ideas, world-views, value systems, art, and education. Hence, authorities determining norms and ideologies, which reinforce the policy platforms in the tourism field, allow dominant groups to act and regulate the social, economic and environmental world about them (Hall, 1996).

A vast proportion of research in tourism and hospitality studies focuses on the normalization, governance in and of tourism, and participation and contestation in decision making of tourism related issues (Baggio *et al.*, 2010; Tosun, 2000; Urry, 1990). However, the concept of people's empowerment for enabling and strengthening their psychological wellness and their general living conditions has more recently attracted attention in academic and professional debates. When power is distributed unequally or a social group is disadvantaged by the official discourse and interactions, striving for empowerment becomes omnipresent. Women, welfare recipients, vulnerable employees, migrants, LGBT groups, patients, elderly people, are some of these groups, which are often considered as achieving some manner of empowerment. While global organizations such as the OECD advocate 'a whole of government' approach in tourism governance, silent voices, which are unable to express themselves, are unobserved in official, academic and professional debates (Isaac *et al.*, 2015; Said, 1978). Relatedly, critical theory researchers have addressed challenging contradictions in tourism studies, including concerns of the marginalized (e.g. Higgins-Desbiolles, 2007; Wearing *et al.*, 2010), power and tourism (e.g. Hall, 2010), critical reflections on participation (e.g. Isaac *et al.*, 2012), contested spaces (e.g. Kousis *et al.*, 2011), and power constellations of global tourism (e.g. Isaac & Çakmak, 2017). These paradoxical matters are picked up again in various chapters in this volume.

Fixed/Singular Identities vis-à-vis Liquid/Plural Aspirations

The fourth paradox in contemporary society and the tourism field is the matter of continually assumed/ascribed singular identities vis-à-vis the actuality of plural and liquid aspirations. Tourism stakeholders could be seen as being at the centre of such matters in many destinations around the world, especially when 'identity' is linked to such concepts as globalization. Over the last century, the prescription of one's identity was largely determined by place of birth and the social status held in the society. The construction of identity was understood to be a matter of human nature, fate, and predestination. In that period people had unambiguous priorities linked to local communities and shared goals, in comparison to the current focus on individualism, self-enlightenment, and self-liberation (Beck & Beck-Gernsheim, 2002).

Today, in the post-modern period, people find themselves with no stable position to aim for in the process of their identity construction (Bauman, 2013; Giddens, 1991). Bauman (2013) terms this formation the 'liquid modernity' that is happening in the 21st century, with significant economic and social change, including increasing flows of people, financial assets, ideas and a requirement for flexibility. As a result, we are moving away from inherited single identities based on place towards hybrid and relational forms of identity characterized by mobility and flux (Easthope, 2009). Individuals thus have to redefine their aims continuously, because if one aims for a particular goal, the likeliness is that not only will the goal posts change over time but also the path one needs to follow to get there (Bauman, 2013). Rather than the linear and simplistic assumptions of the fixed and single identity of yesteryear, then, tourism becomes mixed up with mobility and 'lifestyle' choices and 'projects of the self', so that now 'identity' is never complete and is understood as always fluid in post-modern society (Giddens, 1991). These identity related matters are picked up again in later chapters, and particularly in relation to 'Deleuzian ontologies of belonging and becoming' in Chapter 9 of this volume.

Globally Standardized vis-à-vis Locally Dynamicized

One further major paradox explored in this volume is that of global standardization vis-à-vis local dynamics. While spatial constraints are disappearing, and cultures, people, and information are becoming more interconnected than ever – in order to achieve an increasingly homogenized world – the pressure of 'being competitive' is growing on people, firms, nations, and tourism destinations to reassert their distinctive values and traits. It is paradoxical, therefore, that when individuals encounter some of these homogeneous norms and standards, their response is often to attempt to re-emphasize their own individuality. Globalization is often situated in the current period of postmodernity and represented by the creation of large-scale systems, complex global relations, and built on Wallerstein's 'world system' in particular (Lie, 2003). The hypothesis of globalization is that, from the end of the Cold War in 1991, we have been evolving toward several co-existing economic and political powers in the world, while the distinction between the first, second and third worlds is disappearing (Buell, 1994). The core notion of globalization is that 'we end up with a single human society and therefore necessarily with a world culture' (Wallerstein, 1991: 93–94).

Meanwhile, globalization is often conceptualized as being intrinsically linked to localism. The processes of globalization and localism are connected within organizations at different societal levels. Localism can be seen as the representation of a group's identity by a sense of commitment to a set of cultural practices that are self-consciously articulated and to some degree separated and directed away from the surrounding world (Nadel-Klein, 1991). In that sense localism, articulated across multiple local community levels, becomes highly complex and difficult to interpret. The transformation of the local, or even the personal, context of social experiences, may thus be rooted both in globalization *and in* acts of localism. While the globalization thesis is about structure, an interpretative approach centers on people's daily experiences in the local context (Berger & Luckmann, 1967; Keesing, 1987). The paradox of globalization vis-à-vis localism thus remains in relation to how we understand and interpret grounded and local change affected through tourism, especially in relation to small-scale communities and rural tourism destinations.

Structure of the Book

We have highlighted only briefly here some of the paradoxes constructing, de-constructing, and re-constructing the tourism field and beyond. This volume is intended to provide a contemporary snapshot of the sociological and related thinking about these paradox themes as they arise in tourism contexts and tourism studies. The contents of the chapters are summarised below.

After the editors' introduction to this volume, the following six chapters illustrate different 'scenes' where paradoxes are in evidence. In Chapter 2, Evi Eftychiou focuses on the disputed identity of rural Cyprus and the paradoxes embedded in tourism development discourse there. By focusing on identity politics and tourism in the Troodos mountainous region, this chapter examines the conflict between native elites and rural locals over the definition of modernity. Eftychiou argues that in this post-colonial setting the power of western hegemony not only maintains and legitimizes particular definitions of 'modernity', but also, paradoxically, reverses the definition of 'modern' identity in the cultural setting of Cyprus. In other words, elements that had been classified previously by the native elites as evidence of backwardness were now highly valued, preserved and protected in the name of 'modernity'. Overall, it is argued that the emergence of mass tourism discourse in the 1960s and its transformation into reflexive tourism discourse in the late 1980s is a reproduction of the 'same paradigm', presenting what seems like an absurd contradiction, hence a paradox. However, as explained in this chapter, this could be explained by taking into consideration the *multiple* modernities and traditionalities which exist. In Chapter 3, Emmanuelle Peyvel examines the case of domestic tourism in the mountains of Vietnam and shows how going to the mountains can be a spatial, social and ethnic paradox for the Kinh people. Previously feared, the mountains are now trivialized through tourist experiences in the popular hill stations of the country, and Peyvel argues that this 'paradoxical transgression' provides an opportunity for analysing the articulations between colonial and post-colonial contexts in relation to Vietnam.

Continuing with a similar theme related to evolving heritage and cultural preservation discourses in post-colonial contexts, in Chapter 4, Keith Tan and Paolo Mura discuss the 'logical paradox' of preservation *via* change in relation to the touristic potential of Malaysia's Catholic mission schools. The authors of this chapter explain how a tourism-led change of function for Malaysia's mission schools, which are among the oldest built structures in the country, can be a useful catalyst for the 'preservation' of a historical site's heritage, which would paradoxically be threatened by the 'continuity of use' so often celebrated by tourism scholars and heritage bodies. The 'logical paradox' of this argument hence lies in the fact that 'change' and 'preservation' are opposites, yet to preserve a historical site as a heritage tourist attraction, some degree of change is often necessary.

In Chapter 5, Chen, Burns and J. Wang discuss the Tourism Law of the People's Republic of China and consider the ways in which package tour travelers may be empowered whilst at the same time tourism operators are disempowered. While the Tourism Law was introduced in order to protect and empower package tour travelers, it is argued in this chapter that low-fare package tours in the Chinese context remain in a chaotic state, with various unethical activities, including deception, coercion and even aggression taking place. In addition, the rise of social media has accelerated the circulation of negative reports about low-fare package tours among consumers. The authors of this chapter discuss the ensuing debate on 'over-regulation vs under-regulation', highlighting the paradoxes faced by different stakeholder groups affected by the Tourism Law.

In Chapter 6, Man Tat Cheng discusses cross-cultural encounters in the context of Chinese students' educational visits to London. Asking whether such visits result in sustaining racial prejudice or prompt reflection, Cheng uses this context in order to reflect on the broad assumption that tourism is an instrument to promote peace and mutual understanding. Cheng argues that the paradox exists that this particular tourism sector both reproduces the representation of negative 'racial behaviour', while at the same time challenging the racialization that is prevalently held in Chinese society. Cheng chooses to end the chapter with an encouraging observation whereby the trips offer a valuable opportunity for Chinese students to be reflexive on their experience of building a relationship with black home-stay hosts, thereby challenging their framing of race relations back in their home country.

In Chapter 7, Rose de Vrieze-McBean discusses the contemporary polemics of Chinese outbound tourism to Europe and how China seeks to modify its global image through organizing its outbound tourism and products. This chapter is timely in examining how emerging paradoxes created by Chinese organizations influence the republic's image on the world stage. De Vrieze-McBean argues that the paradox in the republic's outbound tourism operation lays mainly in the effort of blending the four fundamental philosophies, namely, Capitalism, Confucianism, Communism and Consumerism. De Vrieze-McBean uses these four pillars in explaining the characteristics of the Chinese visitors and at the same time in highlighting the impact of Chinese outbound tourism on European destinations, both now and likely in the years to come.

The following two chapters advance thinking about the paradoxes inherent in or encountered in tourism management and development by inspecting some of the key issues involved in contemporary Tourism Studies scholarship. In Chapter 8, Platenkamp discusses paradoxical challenges of the 'internationalized' tourism academia. Addressing the tension between the global and the local in the context of tourism studies, the chapter deals with the linguistic paradoxes in academic discussion. Considering English as the global lingua franca in academic discourse, the question becomes relevant as to how the rich diversity of local languages and cultural traditions might be included, if at all, in order that tourism studies may find a renewed creativity in a rich and diverse, globalizing context. Chapter 9 turns to applying some of the metaphysical insights of the French philosopher Gilles Deleuze (1925–1995) on the arts, film, literature, and science to matters of international tourism. In this chapter, Hollinshead, Suleman, S. Wang, Nair and Vellah seek to translate Deleuze's thinking on non-representational geophilosophy, and find that Deleuze's paradoxical insights have much relevance to 'the creative encounter' possibilities of tourism, especially in terms of how 'space' and 'time' may be rethought as new visions of life unfold around travelers and indeed for resident 'local' populations. The authors argue that the Deleuzean concepts that are offered here attest to the disdain that Deleuze held for most *established* intellectual fashions/*established* disciplines-cum-fields, and, conceivably therefore with relevance to this volume, for the *established* academic realm of Tourism Studies and the operational sphere of Tourism Management. In this chapter, then, the ironies and paradoxes of Deleuzean thought are contextualized within Tourism Studies in terms of what is being inspected and what is not (or not yet) in the amalgam field of Tourism Studies/Tourism Management. The authors of the chapter therefore argue that tourism is a principal site in which Deleuzean ontologies of belonging and becoming could and should be explored.

The editors conclude with Chapter 10, Afterword: Reflections on Paradoxes in Understanding, Culture, Mobility, and Tourism, and provide further thought about the power of travel and tourism in and for our lives. Drawing together and reinforcing the key paradoxical themes through a cogent summary of the various chapters in this book, the editors comment on the wider implications of these paradoxes for both tourism practice and tourism scholarship. Indeed, in embracing, even celebrating, the very dynamic/fluid/messy 'cultural' relationships that exist between tourism, globalization, (post)modernity, and (post)colonialism, a focus on tourism's paradoxical aspects is, as suggested in the philosophical fragment quoted earlier, the 'passion of thought'. The aim of the afterword is therefore to close the book by keeping open our thought lines about the paradoxical propensities of travel and tourism in shaping our identities, in rigging our intensities, and in molding our aspirations, without the chapter – nor the book for that matter – being unduly prescriptive in that end.

References

Alva, J.J.K. de (1995) The postcolonization of the (Latin) American experience: A reconsideration of 'colonialism', 'postcolonialism' and 'mestizaje'. In G. Prakash (ed.) *After Colonialism: Imperialism Histories and Postcolonial Displacements* (pp. 241–275). Princeton, NJ: Princeton University Press.

Ashcroft, B., Griffiths, G. and Tiffin, H. (eds) (1989) *The Empire Writes Back: Theory and Practice in Post-Colonial Literatures*. London: Routledge.

Baggio, R., Scott, N. and Cooper, C. (2010) Network science: A review focused on tourism. *Annals of Tourism Research* 37 (3), 802–827.

Bauman, Z. (2013) *Liquid Modernity*. Chichester, UK: Wiley.

Beck, U. and Beck-Gernsheim, E. (2002) *Individualization*. London: Sage.

Berger, P.L. and Luckmann, T. (1967) *The Social Construction of Reality: Everything that Passes for Knowledge in Society*. London: Allen Lane.

Budde, F., Tranter, P., Fechtel, A., Wise, A., Lui, V. and Milunsky, T. (2013) Winning the next billion Asian travelers – Starting with China. Retrieved 21 November 2018, from https://www.bcgperspectives.com/content/articles/transportation_travel_tourism_globalization_winning_billion_asian_travelers_starting_china/?chapter=3.

Buell, F. (1994) *National Culture and the New Global System*. Baltimore, MD: Johns Hopkins University Press.

Chambers, D. and Buzinde, C. (2015) Tourism and decolonisation: Locating research and self. *Annals of Tourism Research* 51, 1–16.

Chang, T.C. (2015) The Asian wave and critical tourism scholarship. *International Journal of Asia-Pacific Studies* 11, 83–101.

Cohen, E. (1989) 'Primitive and remote: Hill tribe trekking in Thailand'. *Annals of Tourism Research* 16, 30–61.

Cohen, E. and Cohen, S.A. (2015) A mobilities approach to tourism from emerging world regions. *Current Issues in Tourism* 18 (1), 11–43.

Daldal, A. (2014) Power and ideology in Michel Foucault and Antonio Gramsci: A comparative analysis. *Review of History and Political Science* 2 (2), 149–167.

Easthope, H. (2009) Fixed identities in a mobile world? The relationship between mobility, place, and identity. *Identities: Global Studies in Culture and Power* 16 (1), 61–82.

Edensor, T. (1998) *Tourists at the Taj: Performance and Meaning at a Symbolic Site*. London & New York: Routledge.

Foucault, M. (1995) *Discipline and Punish: The Birth of the Prison*. (First published in 1975, translation by Alan Sheridan.) New York: Vintage.

Giddens, A. (1991) *Modernity and Self-identity: Self and Society in the Late Modern Age*. Stanford, CA: Stanford University Press.

Gramsci, A. (1971) *Selections from the Prison Notebooks*. Edited and translated by Quintin Hoare and Geoffrey Nowell Smith. London: Lawrence & Wishart.

Hall, C.M. (1996) Gender and economic interests in tourism prostitution. In Y. Apostolopoulos, S. Leivadi and A. Yiannakis (eds) *The Sociology of Tourism: Theoretical and Empirical Investigations* (pp. 265–280). London: Routledge.

Hall, C.M. (2010) Power in tourism: Tourism in power. In D.V.L. Macleod and J.G. Carrier (eds) *Tourism, Power and Culture: Anthropological Insights* (pp. 199–213). Bristol: Channel View Publications.
Higgins-Desbiolles, F. (2007) Touring the indigenous or transforming consciousness? Reflections on teaching indigenous tourism at university. *The Australian Journal of Indigenous Education* 36 (S1), 108–116.
Hollinshead, K. (2004) Tourism and new sense: Worldmaking and the enunciative value of tourism. In C.M. Hall and H. Tucker (eds) *Tourism and Postcolonialism: Contested Discourses, Identities and Representations* (pp. 25–42). London: Routledge.
Huang, S., van der Veen, R. and Zhang, G. (2014) New era of China tourism research. *Journal of China Tourism Research* 10 (4), 379–387.
Hughes, M., Weaver, D. and Pforr, C. (eds) (2015) *The Practice of Sustainable Tourism: Resolving the Paradox*. Abingdon: Routledge.
Isaac, R. and Çakmak, E. (2017) Exploring the role of science and power relations in tourism studies: An introduction to the special issue. *Tourism, Culture & Communication* 17 (1), 1–6.
Isaac, R.K., Platenkamp, V. and Çakmak, E. (2012) Message from paradise: Critical reflections on the tourism academy in Jerusalem. *Tourism, Culture & Communication* 12 (2), 159–171.
Isaac, R.K., Hall, C.M. and Higgins-Desbiolles, F. (2015) *The Politics and Power of Tourism in Palestine*, Abingdon: Routledge.
Jaakson, R. (2004) Globalisation and neo-colonialist tourism. In C.M. Hall and H. Tucker (eds) *Tourism and Postcolonialism: Contested Discourses, Identities and Representations* (pp. 169–183). London: Routledge.
Jóhannesson, G., Ren, C. and van der Duim, R. (eds) (2015) *Tourism Encounters and Controversies: Ontological Politics of Tourism Development*. Farnham: Ashgate.
Jones, B.G. (ed.) (2006) *Decolonizing International Relations*. Lanham, MD: Rowman & Littlefield.
Keen, D. and Tucker, H. (2012) Future spaces of postcolonialism in tourism. In J. Wilson (ed.) *The Routledge Handbook of Tourism Geographies* (pp. 97–102). London & New York: Routledge.
Keesing, R.M. (1987) Models, 'folk' and 'cultural': Paradigms regained. In D. Holland and N. Quinn (eds) *Cultural Models in Language and Thought* (pp. 369–393). Cambridge: Cambridge University Press.
Kothari, U. (2015) Reworking colonial imaginaries in post-colonial tourist enclaves. *Annals of Tourism Research* 15 (3), 248–266.
Kousis, M., Selwyn, T. and Clark, D. (eds) (2011) *Contested Mediterranean Spaces: Ethnographic Essays in Honour of Charles Tilly*. New York, NY: Berghahn Books.
Lie, R. (2003) *Spaces of Intercultural Communication. An Interdisciplinary Introduction to Communication, Culture, and Globalizing/Localizing Identities*. New York, NY: Hampton Press.
MacCannell, D. (1989) 'Introduction'. *Annals of Tourism Research* 16, 1–6.
Marschall, S. (2004) Commodifying heritage: Post-apartheid monuments and cultural tourism in South Africa. In C.M. Hall and H. Tucker (eds) *Tourism and Postcolonialism: Contested Discourses, Identities and Representations* (pp. 95–112). London: Routledge.
Minca, C. and Oakes, T. (eds) (2006) *Travels in Paradox: Remapping Tourism*. Lanham, MD: Rowman & Littlefield.
Nadel-Klein, J. (1991) Reweaving the fringe: Localism, tradition, and representation in British ethnography. *American Ethnologist* 18 (3), 500–517.
Park, H.Y. (2016) Tourism as reflexive reconstructions of colonial past. *Annals of Tourism Research* 58, 114–127.
Pritchard, A. and Morgan, N. (2007) De-centring tourism's intellectual universe, or traversing the dialogue between change and tradition. In I. Ateljevic, A. Pritchard and N. Morgan (eds) *The Critical Turn in Tourism Studies* (pp. 11–28). Oxford: Elsevier.

Said, E.W. (1978) *Orientalism*. London: Routledge & Kegan Paul.
Soanes, C. and Stevenson, A. (2006) *Oxford Dictionary of English*. 2nd edition revised. Oxford: Oxford University Press.
Thailand Creative and Design Centre (TCDC) (2015) 'My Chiang Mai' exhibition, Thailand, Creative and Design Centre of Chiang Mai.
Tosun, C. (2000) Limits to community participation in the tourism development process in developing countries. *Tourism Management* 21 (6), 613–633.
Tucker, H. and Akama, J. (2009) Tourism as postcolonialism. In T. Jamal and M. Robinson (eds) *The Sage Handbook of Tourism Studies* (pp. 504–520). London: Sage.
Tucker, H. and Zhang, J. (2016) On western-centrism and 'Chineseness' in tourism studies. *Annals of Tourism Research* 61, 250–252.
United Nations World Tourism Organization (2008) UNWTO Tourism Highlights, 2008 Edition. Retrieved 21 November 2018 from https://www.e-unwto.org/doi/pdf/10.18111/9789284419876.
United Nations World Tourism Organization (2018) UNWTO Tourism Highlights, 2018 Edition. Retrieved 21 November 2018 from https://www.e-unwto.org/doi/pdf/10.18111/9789284413560.
Urry, J. (1990) The consumption of tourism. *Sociology* 24 (1), 23–35.
Veijola, S., Molz, J.G., Pyyhtinen, O., Höckert, E. and Grit, A. (eds) (2014) *Disruptive Tourism and its Untidy Guests: Alternative Ontologies for Future Hospitalities*. Basingstoke: Palgrave Macmillan.
Wallerstein, I.M. (1991) *Geopolitics and Geoculture: Essays on the Changing World-system*. Cambridge: Cambridge University Press.
Wearing, S.L., Wearing, M. and McDonald, M. (2010) Understanding local power and interactional processes in sustainable tourism: Exploring village-tour operator relations on the Kokoda Track, Papua New Guinea. *Journal of Sustainable Tourism* 18 (1), 61–76.
Wijesinghe, S.N.R. and Mura, P. (2018) Situating Asian tourism ontologies, epistemologies and methodologies: From colonialism to neo-colonialism. In P. Mura and C. Khoo-Lattimore (eds) *Asian Qualitative Research in Tourism Ontologies, Epistemologies, Methodologies, and Methods* (pp. 97–115). Singapore: Springer Nature.
Winter, T. (2009) Asian tourism and the retreat of Anglo-western centrism in tourism theory. *Current Issues in Tourism* 12 (1), 21–31.
Zhang, J. (2018) How could we be non-western? Some ontological and epistemological ponderings on Chinese tourism research. In P. Mura and C. Khoo-Lattimore (eds) *Asian Qualitative Research in Tourism: Ontologies, Epistemologies, Methodologies, and Methods* (pp. 117–136). Singapore: Springer Nature.

2 The Paradox of Modernity: Power, Identity and Tourism in Rural Cyprus

Evi Eftychiou

Introduction

This chapter is an ethnographic exploration of paradoxes that have emerged in tourism discourse and identity politics in colonial and postcolonial Cyprus. It argues that the power of western hegemony not only defines but also reverses the definition of modernity in Cyprus, and in such a way that its authority is maintained and legitimized. The methodological approach I have adopted is 'global ethnography', as put forth by Burawoy et al. (2000). In the context of tourism discourse, I will be exploring global and transnational phenomena, and their interconnection with local cultural configurations, while historically contextualizing my ethnographical material.

The first part of this chapter deals with those paradoxes I encountered in the field that intrigued me enough to wonder about them. The second part discusses the emergence of tourism discourse in colonial Cyprus, while the third part considers mass tourism as the epitome of modernity on the island in the post-independence period from the 1960s until the 1970s. I focus particularly on the role of western hegemony in defining notions of modernity and development in Cyprus through the lenses of tourism. Building on this, the fourth part examines the changes that occurred in the 1980s and onwards, and the emergence of what I call 'reflexive tourism discourse'. The chapter closes with a detailed discussion of even more paradoxes encountered in the field and explains how these have become part of the conceptual history of tourism in Cyprus and can be best understood within the context of global unequal power relations.

Encountering Paradoxes

The basis of this chapter is a personal story from the field, which I believe reflects the irony and the paradoxes embedded in the cultural history of tourism in Cyprus. In 2004, before I was inspired to start working on this research, I was working for an environmental research unit that had several European projects and organized seminars and conferences with the ultimate goal of informing rural residents in Cyprus about sustainable tourism initiatives. The projects targeted mainly the Troodos mountain range, which consists of a number of villages and is home to the largest national forest and rich biodiversity. For several decades, the area was not included as a priority in the developmental plans of the Cypriot authorities, which focused on the sun and sea, mass tourism model.

In one of our events in the Troodos region, which was attended by several local residents, Dr Andreou (pseudonym), a well-known activist and environmentalist, criticized the current state of mass tourism, stating:

> The tourism that we have today in Cyprus is the tourism of the hotel, the swimming pool and cheap beer ... I dream of the day that I can bring a group of urban children to rural Cyprus, have them stay in a farm house and have the opportunity to pick fruits themselves, to milk the goats, get dirty in the soil and then bathe in the river! But in order to be able to develop this kind of tourism in Troodos, you have to contribute to this effort as well. It is unacceptable for you to come to this beautiful village of ours and be offered a plastic chair to sit down on in the village square! This is unacceptable! This is shameful, damn it! We are a diseased society that appreciates cheap materials over our local products and the labor of our farmers. (Field notes, 2009)

I began to suspect that something was wrong with the sustainable tourism narrative when I noticed the significant cultural gap between the environmentalists' discourse and the rural residents' discourse. My disenchantment with the discourse of sustainability was gradual. During one of my walks in Omodos, a village in the Troodos region, I had a long discussion with Ms Eleni (pseudonym), a woman in her mid-70s. This was the moment I began to feel disillusioned. The heat from the sun was particularly fierce that day, so she invited me to her house for a cold drink. The building was freshly whitewashed and her yard was full of small flowerpots. I sat outside on a white plastic chair and waited for her until she returned with a pitcher of cool lemonade and some homemade cookies. She gave me the impression that she was extremely happy to have someone at her house on a weekday to keep her company until the afternoon church ceremony began. Reflecting on one of the seminars on sustainable tourism, I unconsciously started to identify possible spaces for 'improvement' in the village. I naively asked Ms Eleni why the Community Council hadn't restored a nearby cement-floored alleyway to its earlier stone-paved form, to which she replied as follows:

> Why should we do that, my dear? For us older people, it is much easier walking on cement ... our walking sticks don't get stuck between the

stones. It's dangerous; we may fall and break a leg on these stones. ... It's also cleaner; we throw water on the cement and it's easily cleaned. If you ask younger women, they will tell you the same thing. They can't walk on the stones in high heels. (Field notes, 2010)

Intrigued by Ms Eleni's answer, I decided to probe further. After expressing my agreement with her and admitting that her concerns were well founded, I enquired as to how she felt about the efforts of the Cyprus Tourism Organisation (CTO) to persuade locals to replace their plastic chairs with the old-style wooden ones. Her response was again unexpected:

I don't know what others are doing; the only thing I know is that I am not going to replace them [the chairs]. Are they [the CTO] willing to pay for the wooden chairs? Do they have any clue how much tonenes karekles [handmade, bamboo-bottomed wooden chairs] cost? They are very expensive! It is not only that. ... These plastic chairs will live longer than me! You can leave them outside in the sun, the rain, the dust, and nothing happens to them! What do you think will happen if you leave a wooden chair in the rain and the sun? (Field notes, 2010)

I was astonished by both of Ms Eleni's answers, since I had previously been under the impression that older people at least would be eager to see their village restored to its earlier state by revitalizing the natural environment and local traditions. As a young female urbanite, I nostalgically identified older residents of rural areas as agents of tradition. Ms Eleni proved me wrong. Not only did she refuse to reproduce tradition: she seemed to genuinely enjoy modern amenities and novelties. When our conversation came to an end, she accompanied me to her front door and with a big smile on her face said: 'Since you like our village, you should come and live here, and I will find you a nice young man to settle down with.' I smiled and thanked her, assuming that she had interpreted the fact that I was traveling alone as a sign of being single.

It didn't take me long to realize that my own nostalgic approach to life and my aesthetic inclinations were leading me astray. Nonetheless, the encounter described above was one of several incidents that at once confused and intrigued me, and that in the end convinced me of the worth of this study. I was puzzled by the fact that rural seniors held many traditional ideas, such as a belief in the value of arranged marriages, but that restoring certain aspects of their material tradition, such as the old-style wooden chairs, didn't interest them at all. How had this paradox emerged in Cypriot society? What did these preferences or choices say about the seniors and their relationship with tradition, and with tourism? And what about my own? How was this seemingly minor issue with chairs related to the broader conceptual history of tourism in Cyprus? This chapter is the result of my efforts to find answers to these questions, to frame the local cultural configurations of identity politics within the dominant global discourses of tourism and development.

The Emergence of Tourism Discourse in Colonial Cyprus

British colonization contributed significantly to the development of tourism as a discourse and as a leisure activity in early 20th-century Cyprus. The island had been visited by a few travelers while under Ottoman rule but the number increased significantly with the arrival of the British in 1878, after which Cyprus was considered a 'safe' destination for western travelers. Many authors described Cyprus as 'being backward and underdeveloped' (Persianis, 2007: 29). The ambivalent travel accounts of these authors were filtered through the ideological lenses of western cosmology, which are seen today as 'politicized representations of the colonial legacy in Cyprus' (Eftychiou & Philippou, 2010: 70; Philippou, 2007).

Although the British government recognized the potential of Cyprus as a future tourist destination relatively early on, the limited budget available to promote tourism, together with the prohibitive infrastructure of the island, delayed any actual promotion of tourism until the 1950s. The first British High Commissioner in 1878 predicted that Cyprus' 'pine-clad mountains, with their invigorating air and freedom form malaria, would become a tourist health resort' (Jenness, 1962, cited in Kammas, 1993: 71). British colonial governmental officials were the first to establish tourist resorts at the peak of the Troodos Mountains and in Platres (Ioannides, 1992: 718). These colonial officers were followed by their troops, their families and their camp followers.

Although the British paved the way for the development of tourism in rural Cyprus in the late 19th century, it wasn't until decades later that cultural conditions would allow modern ideas, such as tourism, to take hold among locals. The political, social and economic reforms promoted by the British around the turn of the 20th century, along with the diffusion of western cultural trends in Cyprus, contributed to the development of a new urban culture. From the late 1940s onwards, Cyprus also experienced a period of increasing prosperity and structural changes (Argyrou, 1996: 7). The result of all these changes was that modernity, and all its cultural trappings, was adopted and reproduced by the vast majority of city dwellers (Eftychiou, 2013: 277).

Nonetheless, rural Troodos as a whole remained one of the most disadvantaged areas of the island. The poor living conditions in rural Cyprus didn't allow the vast majority of residents to fully participate in the modernization that was happening in towns. Rural residents experienced town life largely through rare visits and word of mouth. Their preconceived notions of what it meant to be modern were based on the consumption of modernity through its products, such as technology (e.g. television, radio and cars), fashion and architecture.

Regardless of the terrible living conditions in rural Cyprus, six villages in Troodos, the 'hill resorts', attracted the vast majority of tourists to the island, including British colonials and local urban elites. The prosperous visitors, who were perceived by locals as agents of modernity, enjoyed 'modern European' amenities at the hill resorts and specifically in the hotels where they were staying, such as asphalt roads, electricity, running water, and theatrical and musical performances, while the region itself remained mired in poverty. It can be argued, then, that the development of tourism in Troodos' hill resorts acted as an agent of modernity, widening the rift between locals and visitors (Eftychiou, 2013: 104).

There is a general agreement that in the 1950s the colonial government made the first attempt to institutionalize and modernize tourism as a social phenomenon. In 1949, for example, the Tourism Development Office was established in order to 'organize' and 'modernize' the tourist experience, but the measures taken by the colonial office didn't have much time to flower. Although Troodos was identified by the British government as a region with high tourism potential, the British didn't actively invest in that potential. Any attempts that were made were stifled by the political instability during the anti-colonial struggle from 1955 to 1959, and then British rule came to an end when the independent Republic of Cyprus was established in 1960.

Mass Tourism: The Epitome of Modernity in Cyprus

In postcolonial Cyprus, the path towards tourism development and modernization was pursued by the local elites running the newly founded Republic of Cyprus. The narrative and practices employed by the elites centered on the consumption of material goods, technological advancements, a growing infrastructure and economy, entrepreneurship, individualism and competitiveness.

Their chosen narrative was outlined in detail in the official developmental plans of the newly founded state. It was evident that, according to the dominant discourse, nature existed to serve the needs of 'men'. Nature was often, therefore, approached as an obstacle to achieving further development since it was often represented as a 'bottleneck(s) [sic] in the process of optimum economic and social development' (Republic of Cyprus, 1967: 173).

If there is one basic principle that is embedded in the development plans of the new republic in the 1960s and 1970s, it is the 'necessity' or fetishism of 'rapid and unhindered development'. The idea that Cyprus was 'left behind' and needed to quickly 'catch up' with other European nations is reflected in almost all sections of the development plans

(Republic of Cyprus, 1967: 209). As a result, the state played a significant role in channeling investments and giving priority to regions that the local elites identified as 'areas with tourism potential' (Republic of Cyprus, 1961: 13), namely, the seaside resorts of Famagusta and Kyrenia, and later on Limassol, Larnaca and Paphos (Republic of Cyprus, 1972: 210).

The elites were very quick both to internalize and to capitalize on the cultural change in western societies that linked tourism with sun bathing (Republic of Cyprus, 1967: 207). As a result, tourism authorities and brokers promoted Cyprus as a 'Mediterranean Sea and sun destination'. In the postcolonial setting, the beach was transformed from a downgraded, unwanted, unproductive space to a symbol of modernity, cosmopolitanism, progressiveness and individual liberation.

The postcard reproduced reflects the growth fetishism and the neo-functionalist approach to development adopted by the Cypriot authorities (Attalides, 1993: 218). The result of the developmental policies enforced during the 1960s and 1970s was a 900% increase in tourist arrivals. Although tourist arrivals jumped from 25,000 in the 1960s to more than a quarter a million in 1973 (Witt, 1991: 37), the third developmental plan, published in 1972, still stuck to the objective of 'increasing numbers of foreign tourists and excursionists' (Republic of Cyprus, 1972: 208).

It wasn't only the local elites who were altering the face of tourism in Cyprus, however. Large-scale tourism in Cyprus can also be linked to the western vision of development, as outlined by the modernist paradigm. In the 20th century, countries were generally divided into three categories: underdeveloped, developing and developed (Argyrou, 2005: 27). Cypriots saw themselves as belonging to the second category, which affirmed their supposed separation from so-called developed countries. The aim of the Cypriot elites was, therefore, to 'develop' Cyprus as a society so that they could join the third category and become fully developed and modern.

The modernist paradigm, as constructed in the 19th century, is Eurocentric and derives its authority from the superior status northern Europeans enjoyed in the world (Argyrou, 2005: 15; Isaac, 2016; Wallerstein, 2006). Europeans predicated their power on their knowledge and ability to master or subordinate nature to their own needs using science (Argyrou, 2005: 11). The legitimacy of such a paradigm was never questioned. European scientists and institutions drafted reports and guidelines to be followed by the 'backward' and 'traditional' countries, which would show them how to 'modernize' so they could develop economically (Argyrou, 2005: 28; Escobar, 1991: 663). The idea was simple and straightforward: 'developing' countries should utilize western know-how to leap across the centuries and transform their 'traditional', 'backward', 'underdeveloped' countries into 'modern', 'developed' nation states (Argyrou, 2005: 33).

The internalization of the modernization and development discourse by the rest of the world produced a new world order: *western hegemony*

(Argyrou, 2005: 22–27). Western hegemony should not be dismissed as simply another school of thought or another popular ideology. The main difference between the two concepts is that ideologies are conscious, meaning that a person is aware of alternative ideas that exist out there but 'consciously' accepts some and rejects others. Hegemony, on the other hand, is much more powerful: it produces specific cultural conditions in which alternative visions are unthought, unnatural or irrational (Comaroff & Comaroff, 1991: 23). As Escobar (2011[1995]: 5) argues: 'reality, in sum, had been colonized by the development discourse' as it emerged in the west.

It seems that postcolonial Cyprus was not an exception. I suggest that western hegemony defined the content of Cypriot identity and locals' vision of tourism development. The western rhetoric of 'modernization and development' dominated the discourse of the newly formed Republic of Cyprus. Local elites, most of whom were educated in the United Kingdom, France, Germany, Greece or Turkey, internalized the assumption that they could rapidly transform their homeland into a modernized, developed tourist destination. Such was the power of the western model of development that, after Cypriots gained their independence, there was no public debate about who they were, where they belonged, where they were going and what their vision for their country should be. As Bauman (1996: 19, cited in Tilley, 2006: 11) argues, identity questions are born of uncertainty. In the case of postcolonial Cyprus, it seems that local elites were certain about the superiority of European civilization and its technological and scientific achievements. In short, Cypriots' 'truth' was already defined and constrained by the power of western hegemony (Foucault, 1980: 131).

Turning now to the development vision promoted by the elites for the villages in the wider Troodos region, I suggest that this, too, did not escape from the modernist paradigm. Rural areas were seen as disadvantaged and backward places inhabited by peasants. As a result, the elites drafting the strategic plans for the state identified the need to modernize rural Cyprus and provide the opportunity for rural communities to 'participate in the social advantages of a modern society' (Republic of Cyprus, 1967: 24).

Consider, for example, rural houses under the plan. Rural residents were viewed as disadvantaged individuals because they did not enjoy the modern amenities available in the cities. Housing conditions were one of the issues that, according to the government, needed 'urgent' attention. The elites perceived old houses, i.e. those built in the 1930s or 1940s, as symbols of a non-modern and backward society, and they felt that this had to change (Republic of Cyprus, 1967: 169). One of the ways in which the government sought to improve existing housing conditions was by encouraging some form of mass production and standardization in construction (Meethan, 2001: 21; Republic of Cyprus, 1967: 172).

It is clear, therefore, that according to the plans for the new republic, modernity was desired not only for urban centers but also for rural areas, which were treated as embarrassing signifiers of an unwanted past.

The result of the prolonged negligence of Troodos by the Cypriot authorities resulted in an enormous cultural gap dividing urban centres from rural areas. In order to appreciate this point, consider the developmental paradox that existed in Cyprus in the 1960s: in Famagusta, tourists and locals enjoyed amenities such as storage heaters, fans and hot showers, and entertainment that included fashion shows, beauty contests and water sports, whereas in the island's rural areas, running water and electricity were still issues. In the same period, for example, there were 200 villages without access to electricity, and 33% of the rural population had no running water in their homes (Republic of Cyprus, 1967: 128, 162).

Although the road to large-scale tourism development in Cyprus was interrupted for a while by the Turkish invasion in 1974, the so-called economic miracle of the late 1970s and 1980s gave rural residents the opportunity to show the world their modern identity and lifestyle. Local authorities and the vast majority of Troodos' residents were actively involved in the process of modernizing the cultural landscape. Among other things, old houses were demolished and replaced by modern ones; high-rise buildings and houses were constructed; trees were removed and mountains were cut into for the creation of car parks, and stone-paved streets were asphalted.

The Change: The Emergence of the 'Reflexive Tourism' Discourse

In the meantime, in the 1980s, just as rural residents had begun to enjoy the material fruits of modernity, local elites were engaged in the process of reversing its very definition by following a new paradigm that had emerged in the western world. More specifically, the majority of intellectuals, local experts and active citizens began to adopt and (re)produce the new discourse of 'modernity' and 'development' that had emerged in the 1960s in the so-called more developed countries of the west (Argyrou, 2005: 39; Byrne, 1991; Escobar, 2011 [1995]: 196; Macnaghten & Urry, 1998). The dominant rhetoric and narrative was now focused on the urgent 'need' to develop sustainably so as to preserve the environment and heritage of Cyprus. The new western narrative on environment, heritage and development was adopted part and parcel by elites in Cyprus and then bulwarked by new laws and internationally funded programs (Smith, 2006: 21).

By the 1990s, the official rhetoric of the island's national authorities came to terms with the dominant discourse promoted by the elites. The rhetoric of the 1960s, which supported the rapid development of mass, packaged and standardized tourism in coastal areas, was transformed in the late 1980s into the discourse of what I call 'reflexive tourism', inspired by Beck (1992 [1986]), Giddens *et al.* (1994) and Welz (2000), which advocates small-scale, unregulated, high-quality, culturally immersive experiences in a rural environment undertaken in a sustainable and responsible manner (Cyprus Tourism Organisation, 2000: 12; Republic of Cyprus, 1989: 156). A few years later, the Cyprus Tourism Organisation's (CTO) *Strategic Plan for Tourism 2000–2010* and its *Strategic Plan for Tourism 2011–2015* declared 'sustainable development' a major priority (Cyprus Tourism Organisation, 2000: 12; Cyprus Tourism Organisation, 2011: 15).

In the context of reflexive tourism, nature was reconceptualized as a 'fragile environment' that in its 'pure' form was considered an 'asset' to the tourism industry (Republic of Cyprus, 1994; Republic of Cyprus, 1999: 423–431; Republic of Cyprus, 2007: 92). This idea has persisted until the present. Rural landscapes such as Troodos have been romantically recast as 'nature's miracle', to be consumed by gazing. Similarly, tradition, which was once associated with backwardness, has been transformed into cultural heritage, which, according to local elites, should be protected, preserved, conserved and, above all, organized and displayed for tourist consumption. The 'underdeveloped' region of Troodos is now identified as the ideal place to implement environment and heritage conservation projects, with the ultimate goal of developing small-scale, cultural tourism in the area. Seen in this light, local elites have re-appropriated material tradition. In other words, elements that were once classified as evidence of backwardness, such as old houses and stone-paved streets, have been recast as fragile objects to be preserved and protected. While a small minority of young people and tourism brokers in Troodos have adopted the discourse of reflexive tourism and share a strong belief that this is the way forward, the majority of Troodos' residents are less willing to reproduce modernity's new discourse, which leaves them exposed to the gazing eyes of the local elites and experts.

Encountering Paradoxes

The paradoxes addressed in this chapter are linked with the ambivalence of postcolonialism as a discourse and as a context (Hall & Tucker, 2004). The ethnographic data discussed below indicate why colonialism and hegemony are relevant to tourism as a discourse and social phenomenon. The first paradox I would like to discuss is the paradoxical simultaneous acceptance and rejection of European superiority. It should not be assumed that the elites' acceptance of western ideas on development and modernization went completely unchallenged. On the contrary, this research suggests that Cypriots partly challenged the intellectual and spiritual authority of modern European identity. As Argyrou (2005) mentions in the postcolonial contexts of India and Africa, where there was also a paradoxical acceptance and rejection of European superiority, while local populations recognized the contribution of European civilization to material goods in the realms of the economy, technology and science, they rejected what Europe had to offer on a spiritual level (July, 1968: 476, cited in Argyrou, 2005: 22; Chatterjee, 1993: 6).

In the case of Cyprus, the material aspects of western civilization were not only acknowledged as superior, they were coveted (Markides *et al.*, 1978: 205). There were no objections raised against adopting western technological advancements, such as electricity, water supply systems, road and building construction, television, radio, and so many other modern amenities that Europeans widely enjoyed. As far as the intellectual and spiritual aspects of modernism were concerned, however, locals did raise objections, but mainly in the areas of religion, family values and gender roles.

A common criticism leveled against modernity in Cyprus, for example, focused on the political emancipation of women and their changing roles in family and society. In the 1960s, being characterized as a 'modern woman' was offensive to Cypriot women themselves, since it implied immodest behavior. The achievements of 'modern women' were mockingly equated by local newspapers with undisciplined behavior, improper dressing and extramarital sexual relations (Monternismoi, 1969: 3). In rural settings, such behavior was also perceived as an imitation of the 'urban way of life', and women who adopted it were isolated and harshly criticized for being 'shameless' (Markides *et al.*, 1978: 89–90). In Platres, one of the hill resorts of the Troodos Mountains, the priest of the village used to ban what he considered urban, lipstick-wearing women from his church because their behavior was improper. It seems, then, that 'modern ideas' were not exactly embraced by rural communities in the Troodos region. It is important to note here that resistance was largely focused on the ideological aspects of modernity and not so much on its material expressions. In the case of the angry priest, his resistance expressed towards women with lipstick can be interpreted as resistance towards the symbolic meaning of lipstick, which here represents the western trend of female emancipation.

Rural residents of Troodos might have come a long way since the 1960s, and lipstick might not be an issue any more, but I argue that there is still a paradox at the heart of the villagers' relationship with modernity. While they accept the material outcomes of modernity, they continue to reject the intellectual ramifications of the process, such as gender equality. One of my fieldwork encounters illustrates this paradox perfectly.

Kakopetria is one of the hill resorts located in the Troodos region that was earmarked by local authorities in the 1980s and 1990s as one of the traditional architectural emblems of Cyprus worthy of preservation. Kakopetria is divided into an old and new neighborhood (Figure 2.1). In the 1960s and 1970s, only a small number of elderly residents were still

Figure 2.1 Old and new Kakopetria village (author photograph)

living in the old neighborhood. Mrs Stavroula (a pseudonym) who had lived in old Kakopetria since 1963, described the neighborhood as it had been before the 1970s:

> It was a big mess, very dirty. Instead of this stone-paved street that you see now, it was a dirt road with no sewage system. People used to throw dirty water from balconies of their houses, there were chickens running up and down ... There was an old man who could not move easily and he used to empty his chamber pot off his balcony. It was so cheap to buy a house here, only 100 pounds. People who had money used to buy big fields in the new village and they could build their house as big as they wanted. (Recorded)

Following Bourdieu (2008 [1979]), the new neighborhood gave wealthier people the opportunity to distinguish themselves from their 'inferior' co-villagers, establishing a division within the village. Most of the new houses were of a similar style: multi-storied, imposing, modern structures with modern amenities. The house was now perceived as a signifier of social status, prestige and identity. The desire to build multi-storied houses that were bigger and more ostentatious was common in many of the region's villages at the time (Markides et al., 1978: 78).

Nowadays, old Kakopetria is under the protection of the Department of Antiquities and the Department of Town Planning and Housing, which continue the work of restoring old Kakopetria to how it was in the pre-modern period and thus converting it into a tourist attraction. Locals and owners of houses in the old neighborhood constantly complain about the strict design and aesthetic codes imposed by the 'Museum', while archaeologists and architects, whose ultimate goal is to purify, ruralize and traditionalize the village, participate actively in what Urry (2002 [1990]: 120) has described as 'designing for the gaze'.

One of the ways they do this is by removing all elements and materials associated with modernity and novelty and replacing them with 'natural' or 'traditional' materials. For example, aluminum window frames have been replaced with wooden ones, asphalt roads with stone-paved roads, iron railings with wooden railings, and plastic chairs with wooden chairs. Furthermore, all reminders of modernity, including air conditioner superchargers, central heating boilers, solar panels, television antennae and satellite dishes, that were seen to be polluting the purity of the area, have been removed or hidden from sight (Figure 2.2). In this context of strategic traditionalization and purification of the rural scenery, Ms Eleni, whose story I told at the beginning of this chapter, would have a difficult time convincing the experts of the legitimate use of her beloved white, plastic chairs.

The 'inspecting gaze' of the museum representatives eventually became part of the daily routine of locals, especially during the 1990s. Some of the locals were dragged to court for refusing to make certain changes or for not complying fully with the requirements of the authorities. The vast majority of those who inherited a house in old Kakopetria expressed strong frustration and anger towards the 'Museum'. The following excerpt is from a discussion I had with Andreas (a pseudonym), a 50-year-old resident of Kakopetria and his wife Maria (pseudonym), about the 'inspecting gaze' of the archaeologists:

A: When we were restoring the house, we did not ask for any money from the Museum; we did everything by ourselves. But we were unlucky because the Museum told us that our house was the oldest house in Old Kakopetria, so they took the initiative and declared the building an ancient monument. They used to come often and check up on the restoration of the house. Every time they came, we used to fight with them. One time, that lady from the Museum [archaeologist] came, and she told me I had to restore the old door and put it back on the house. I told her, 'The door is very low, and we will hit our heads on the lintel upon entering the house.' It was only one meter and forty centimeters tall! Look at me! I'm two meters tall! (My emphasis.)

E: Why was the door initially so low?

A: Because in the past the lower level of a house was used for animals. So, basically, the opening was high enough for animals.

Figure 2.2 A building in old Kakopetria, showing the alterations that owners have made (author photograph)

E: What happened later? Did she accept the alteration?
A: No! She insisted that we restore the old door and put it back on the house. Then I asked, 'How will we enter the house?' She made a mistake, and she told me that we should bend upon entering! That was my chance to get back at her, and I replied, 'Listen to me: only women bend over, not men!'

(Recorded)

Andreas was very proud to share his story with me. He laughed when he repeated the part where he made the knowingly sexist comment to the archaeologist. When he left to go to the coffee shop, Maria, his wife, looked to the left and right and then told me in a very soft voice, as if she was afraid to mention it, that after that wrangle with the archaeologist, they received a court summons. 'We couldn't do anything else,' she explained. 'They would have forced us to do it anyway. I don't regret it.'

It was evident that traditional gender roles were still alive and healthy in Mr Andreas' household. Andreas was the alpha male of the house, making all decisions singlehandedly, and Maria felt that she needed to support all those decisions; criticism wasn't tolerated. The fact that the Department of Antiquities, and even worse, a woman representing the 'Museum', had the authority and the power to tell Andreas what to do in his own house was not something that he was willing to tolerate. Defending his right to make decisions on his property was evidently linked to defending his masculinity; hence, his outburst.

It is important to note that the process of normalizing the local elites' discourse is complex and intertwined with global and local power struggles. At the local level, rural residents, in their efforts to change in the face of the hegemonic positions of the elites, resist the aesthetic traditionalization and seemingly arbitrary regulations imposed on them. This intransigence has put local residents in an ironic position. As mentioned earlier, their refusal to acquiesce to the new paradigm of development and their persistence in striving for material modernity left them once again exposed as 'backward', 'ignorant', 'parvenu' peasants.

From this complex interplay of discourses and binary divisions, another paradox emerges. In the 1980s, the notion of modernity was redefined by local elites as the development of small-scale, unpackaged, rural tourism oriented towards the protection of the 'environment' and 'heritage'. Paradoxically, according to this new understanding of modernity, if one wants to be perceived as 'modern', one needs to respect tradition. But the material expressions of tradition are not necessarily aligned with its intellectual aspects. For example, a 'modern' elite family in Cyprus might be seek out a house with traditional architecture, cook traditional recipes and decorate their home with traditional art and crafts, without, however, adopting traditional values like gender division. Emilios (a pseudonym), an architect in training and the CTO officer responsible for Cyprus' agrotourism program, explained to me how he sees himself as 'modern' because he was greatly influenced by the modernist paradigm, while recognizing the critiques of modernity he was exposed to in the final years of his studies in the UK:

> I carry with me a sensitivity towards the environment, long-term planning and design. I am forced to have a good connection with nature, land ... I refer to tradition often with a romantic and nostalgic approach to the various types and expressions of tradition. What is really important,

though, is that tradition keeps me well connected with the past ... there are several values [of tradition] that we choose to leave behind, and we preserve the 'appearance' of tradition ... there are so many things [traditional values] that I had to get rid of. For example, gender inequality or the dowry system – all these mentalities and behaviors that are linked with another society, not the contemporary one. (Recorded)

Emilios considers himself a child of modernity. While he appreciates 'gazing upon' and 'consuming' tradition, he dislikes the idea that tradition should have the power to dictate aspects of his personal life. While respecting material traditions, he strategically rejects 'traditional' ways of thinking, such as male chauvinism.

Conclusions

Thinking of tourism and identity politics in terms of discourse, reveals the power relations involved in the constantly evolving process of identify formation and the ways in which western hegemony has managed to dominate that discourse in less powerful societies such as Cyprus. My main argument is that the power of western hegemony has not only defined but also reversed the definition of modernity in Cyprus, and in such a way that its authority has been maintained and legitimized.

The concepts of western hegemony and modernization have been used here in a Foucauldian sense. Power relations have, thus, been identified within the west itself. Specifically, modernity is approached as a paradigm by which the powerful groups in western societies differentiated themselves from the less privileged; the dominated groups of the west – 'the Other within' (Argyrou, 2005). Hence, modernization can be seen as the effort of dominant groups in the west to 'universalize their culture' both within the west itself and throughout the rest of the world. As Wallerstein (2006) puts it, even dominant actors need legitimization for their advantages and privileges, and the rhetoric of modernity creates the ideological framework in which legitimization of intervention takes place.

Cypriots, like many other colonized people, have embraced the idea that western culture is superior to their own and thus endeavor to achieve modernity using the 'more advanced' countries of the west as a benchmark. The local elites' 'truth' is already defined and constrained by the power of western hegemony to the degree that the western model is considered the 'natural' and 'rational' way forward, even as it has reversed itself over time.

As a rule, more powerful cultures have the authority to define the 'socio-historical unconscious' (Argyrou, 2005). It seems that for now, the west holds the power to define the dominant discourse, according to which western civilization is superior to others. This hegemony is extremely difficult to challenge by virtue of what Foucault calls the 'process of division' (Rabinow, 1991 [1984]: 8). Through this process, the groups who have the power to define concepts objectify and categorize the world around them based on binary systems of thought, such as west/east, First World/Third World, modern/traditional, developed/underdeveloped, progressive/backward and urban/rural. The world acquires meaning, and social attitudes are regulated according to these classifications alone. Following Foucault, I claim that Cypriots have positioned themselves as subjects of western hegemony by adopting and reproducing the discourse of modernity and development as it evolved in the west.

The ethnographic material presented in this chapter reveals that paradoxes are embedded in the contested process of defining, negotiating and re-negotiating our own identities and the identities of others. Such paradoxes serve to remind us of Said's (1996 [1978]) work and the dangers of 'essentializing' the conceptual categories that we use in order to make sense of ourselves and the Other. The fact that I, as a researcher, identified as paradoxical the behavior of Ms Eleni, an old lady who holds traditional values but deeply appreciates modern amenities, reveals as much about me as about her. The anthropological gaze is an equally organized and systematized way of looking at things in a constructed context. Paradoxes are seen as 'paradoxes' only when the anthropological gaze has a preconceived notion of how things 'should' be. For this reason, as scholars, we need to acknowledge the multiplicity of constantly shifting analytical categories at our disposal and, following Bourdieu (1990), challenge the binary categories embedded in western epistemology.

References

Argyrou, V. (1996) *Tradition and Modernity in the Mediterranean: The Wedding as a Symbolic Struggle*. Cambridge: Cambridge University Press.
Argyrou, V. (2005) *The Logic of Environmentalism: Anthropology, Ecology and Postcoloniality*. New York & Oxford: Berghahn Books.
Attalides, M. (1993) Paragontes pou Diamorfosan tin Kupriaki Koinonia meta tin Aneksartisia [Factors that shaped Cyprus's society after independence]. In *Kypriaki Zoi kai Koinonia: Ligo Prin tin Aneksartisia kai Mexri to 1984* [Cyprus's life and society: From just before Independence until 1984], lectures at Open University, Nicosia: Municipality of Nicosia.
Bauman, Z. (1996) From pilgrim to tourist, or a short history of identity. In S. Hall and P.D. Gay (eds) *Questions of Cultural Identity* (pp.18–36). London: Sage.
Beck, U. (1992) [1986] *Risk Society: Towards a New Modernity*. London: Sage.
Bourdieu, P. (1990) *The Logic of Practice*. Palo Alto, CA: Stanford University Press.
Bourdieu, P. (2008) [1979] *Distinction: A Social Critique of the Judgement of Taste*. New York & London: Routledge.
Burawoy, M., Blum, J.A., George S., Gille Z., Gowan T., Haney L., Klawitter M., Lopez S.H., O'Riain, S. and Thayer, M. (2000) *Global Ethnography: Forces, Connections and Imaginations in a Postmodern World*. London: California University Press.
Byrne, D. (1991) Western hegemony in archaeological heritage management. *History and Anthropology* 5 (2), 269–76.

Chatterjee, P. (1993) *The Nation and its Fragments: Colonial and Postcolonial Histories*. Princeton, NJ: Princeton University Press.
Comaroff, J. and Comaroff, J. (1991) *Of Revelation and Revolution: Christianity, Colonialism and Consciousness in South Africa, Vol. 1*. Chicago, IL: Chicago University Press.
Cyprus Tourism Organisation (2000) *Strategic Plan for Tourism 2000–2010*.
Cyprus Tourism Organisation (2011) *Strategic Plan for Tourism 2011–2015*.
Department of Forests (2011) *The State of the World's Forest Genetic Resources, Country Report: Cyprus*. Nicosia, Cyprus: Department of Forests, Ministry of Agriculture, Natural Resources and Environment,
Eftychiou, E. (2013) Power and tourism: Negotiating identity in rural Cyprus. PhD thesis, University of Hull.
Eftychiou, E. and Philippou, N. (2010) Coffee-house culture and tourism in Cyprus: A traditionalized experience. In L. Jolliffe (ed.) *Coffee Culture, Destinations and Tourism* (pp. 66–86). Bristol: Channel View Publications.
Escobar, A. (1991) Anthropology and the development encounter: The making and marketing of development anthropology. *American Ethnologist* 18 (4), 658–682.
Escobar, A. (2011) [1995] *Encountering Development: The Making and Unmaking of the Third World*. Princeton, NJ: Princeton University Press.
Foucault, M. (1980) *Power/Knowledge: Selected Interviews and Other Writings 1972* (edited by C. Gordon). London: Longman.
Foucault, M. (1984) What is enlightenment? In P. Rabinow (ed.) *The Foucault Reader* (pp. 32–51). New York, NY: Pantheon.
Giddens, A., Beck, U. and Lash, S. (1994) *Reflexive Modernization: Politics, Tradition and Aesthetics in the Modern Social Order*. Cambridge: Polity Press.
Hall, C.M. and Tucker, H. (eds) (2004) *Tourism and Postcolonialism: Contested Discourses, Identities and Representations*. New York & London: Routledge.
Ioannides, D. (1992) Tourism development agents: The Cypriot resort cycle. *Annals of Tourism Research* 19, 711–31.
Isaac, R.K. (2016) Eurocentrism tourism. In J. Jafari and H. Xiao (eds) *Encyclopedia of Tourism* (pp. 1–2). New York, NY: Springer.
Jenness, D. (1962) *The Economy of Cyprus: A Survey to 1914*. Montreal: McGill University Press.
July, R.W. (1968) *The Origins of Modern African Thought*. London: Faber & Faber.
Kammas, M. (1993) The positive and negative effects of tourism development in Cyprus. *Cyprus Review* 5 (1), 70–89.
Macnaghten, P. and Urry, J. (1998) *Contested Natures*. London: Sage.
Markides, K.C., Nikita, E. and Rangou, E.N. (1978) *Lysi: Social Change in a Cypriot Village*. Nicosia: Publications of the Social Research Center.
Meethan, K. (2001) *Tourism in Global Society: Place, Culture, Consumption*. Basingstoke: Palgrave.
Monternismoi. (Modernization). (1969) Maxi, 3 September, p. 3.
Persianis, K.P. (2007) *Polis kai Politismos: O Rolos ton Kipriakon Poleon sti Dimiourgia tou Neoterou Politismou tis Kiprou 1878-1931* [Cities and Civilization: The Role of Cypriot Towns in the Creation of Modern Civilization in Cyprus 1878–1931]. Nicosia: Intercollege Press.
Philippou, N. (2007) *Coffee House Embellishments*. Nicosia: University of Nicosia Press.
Rabinow, P. (1991) [1984] *The Foucault Reader: An introduction to Foucault's Thought*. London: Penguin Books.
Republic of Cyprus (1961) *The First Five-Year Plan of Economic Development*. Nicosia: Planning Bureau, Republic of Cyprus.
Republic of Cyprus (1967) *The Second Five-Year Plan 1967–1971*. Nicosia: Planning Bureau, Republic of Cyprus.
Republic of Cyprus (1972) *The Third Five-Year Plan 1972–1976*. Nicosia: Planning Bureau, Republic of Cyprus.

Republic of Cyprus (1989) *Five-Year Development Plan 1989–1993*. Nicosia: Planning Bureau, Republic of Cyprus.
Said, E.W. (1996) [1978] *Orientalism* (Οριενταλισμς) translated into Greek by F. Terzakis. Athens: Nefeli Publications.
Smith, L. (2006) *Use of Heritage*. London & New York: Routledge.
Tilley, C. (2006) Introduction: Identity, place, landscape and heritage. *Journal of Material Culture* 11 (1–2), 7–32.
Urry, J. (2002) [1990] *The Tourist Gaze*. London: Sage.
Wallerstein, I. (2006) *European Universalism: The Rhetoric of Power*. New York & London: The New Press.
Welz, G. (2000) Multiple modernities and reflexive traditionalisation: A Mediterranean case study. *Ethnologia Europaea* 30, 5–14.
Witt, F. (1991) Tourism in Cyprus: Balancing the benefits and costs. *Tourism Management* 12, 37–46.

3 Go West! Overcoming the Paradoxes of Kinh Tourism in the Vietnamese Mountains: A Postcolonial Geography

Emmanuelle Peyvel

> Life has just begun to burst forth.
> I want to seize the clouds and wind,
> Drunk with love on butterfly wings.
> I want to embrace in an ardent kiss
> The mountains, streams, trees, and bright grass
> To delight in this world of perfume and light,
> To satiate my soul with the prime of life.
> O, vermeil spring! I want to bite into thee!
>
> Xuân Diệu, Ta Muốn Ôm [I want to grasp], 1938

Introduction

While Vietnam welcomed more than 18 million international tourists in 2019, 85 million Vietnamese were tourists in their own country. Domestic tourism is a stronghold of this economic sector, which has gone through extreme growth, increasing by a multiple of eight in 20 years (Vietnam National Administration of Tourism, 2019). Vietnamese tourists travel preferentially to seaside resorts – especially Vũng Tàu in the south, Phan Thiết and Nha Trang in the center, Sầm Sơn and Đồ Sơn in the north, as well as islands such as Phú Quốc and Côn Đảo in the south. There are also many who visit UNESCO world heritage sites – in particular Hạ Long Bay, the imperial city of Huế, Mỹ Sơn and Hội An, and the two main cities of the country: Hà Nội, the political capital in the north and Hồ Chí Minh City, the economic capital in the south.

Those destinations are structured by the triptych 'Pray, pay and play' (Graburn, 1983a), that sums up quite well Vietnamese practices: going to pagodas, visiting national heritage, shopping, celebrating and bathing.

Mountains have only just begun to be involved in this triptych. Indeed, there are no major touristic sites in the Vietnamese mountains. The tourism economy is rather on the eastern side of the country, along the coast, in the rice-farming lowlands and deltas. In a country where mountains cover approximately two-thirds of the territory, this situation appears as a paradox. Indeed, for Kinh people, altitude is a 'new reality' (Papin, 2003), only invested in for domestic tourism purposes during the late 20th century. This dominant ethnic group, which represents 86% of the national population (General Statistics Office of Vietnam, 2016), is known for living in the eastern rice-growing deltas and plains, that is to say the wealthiest regions of the country that are nowadays increasingly urbanized. By contrast, the western mountains have long been unvalued regions, left aside for the 53 other ethnic groups. This vertical partition of the land is still today the backbone of Vietnam's spatial organization and firmly anchored in people's representations and behaviors. This is one of the main reasons Kinh people experience alterity while engaging in tourism in the mountains. This spatial segmentation is also typical of numerous Asian countries, such as China, Thailand or Laos (Bruneau et al., 1995; Koninck, 2016; Scott, 2009). In Vietnam, as in these other countries, this demographic and spatial imbalance also translates into an unequal share of political and economic power: while the Kinh occupy most positions of power, the 53 other ethnic groups are designated as inferior minorities (dân tộc thiểu số). From 1986, the growth that occurred following the renovation policy which opened the country to the market economy (Đổi Mới) primarily benefited the Kinh and their territories. Since 2000, this economic growth has exceeded a yearly average increase of 6%, substantially improving the living standards for this ethnic group and yielding unprecedented access to tourism.

For the Kinh people, engaging in tourism in the mountains could be said to be a triple paradox. The first is spatial: leaving the comfort of home to reach higher, more peripheral regions. While the urban lowlands have embraced global consumerism since the Đổi Mới, mountains are less connected to global commodities. The second is socio-ethnic: historically, the mountains are not Kinh territory and they are also poorer regions possessing less infrastructure dedicated to tourism. The third is representational: mountains are feared environments that have mostly been depreciated in Kinh culture. Indeed, folk tales and legends depict mountains as dangerous areas, home to savage beasts, monsters and bad spirits. Thus, we might expect higher altitudes to be no-go zones for the Kinh. However, in recent years, mountains have become the new frontier, a paradoxical expansion of what, as already mentioned, Nelson Graburn describes as the 'pray, play and pay' triptych. This chapter aims at unpacking the paradox of the Kinh's mountainous expansion, what this reveals, and the limitations it includes.

Beyond the Paradoxes: Engaging with Postcolonial Geography

In a chronological reading, mountains could appear as a new playground that was once ignored and now rediscovered. In order to go beyond this simplistic view, a geographical perspective allows one to probe into the articulations between colonial and postcolonial contexts (Hall & Tucker, 2004; Tucker & Akama, 2009) linking places, practices and representations of mountains. The postcolonial lens helps in reassessing the colonial period as an intense moment of spatial production, with the making of dedicated sites such as four hill stations in Vietnam, and the rise of a way of thought that concurrently valued and devalued mountain spaces and populations according to an exoticist and primitivist schema (Said, 1978). These material and immaterial legacies of the colonial period still bear currency in the sense that they have been reclaimed by the dominant ethnic group – the Kinh – who not only frequent those hill stations but also mix in new and old practices, such as pilgrimage.

This hybridization process is an opportunity to extend Hall and Tucker's inquiry on the following paradox: 'Why postcolonial societies should continue to engage with the imperial experience, since nearly all postcolonial societies have achieved political independence' (Hall & Tucker, 2004: 4). However, rather than looking at international tourism and dedicated products such as safaris, this study provides a new perspective by focusing on domestic tourists of a former colony reclaiming spaces they were excluded from. For a Kinh tourist in a postcolonial context, doing mountain tourism involves not only reclaiming colonial architecture and landscape, but also using the mountain frame to reinstate a dominant position within the Vietnamese ethnoscape. These nested logics of dominance are a terrain for understanding how colonial ways of thought, such as exoticism and primitivism, may abide among formerly colonized populations, and legitimize paradoxical transgressions.

More generally, the study of domestic tourism in a former French colony like Vietnam permits more broadly a decentering of tourism studies (Ghimire, 2001; Graburn, 1983b; Hitchock et al., 1993; Nash, 1981; Singh, 2009; Teo et al., 2001; Winter et al., 2009). Indeed, this case study also constitutes an opportunity to better understand the globalization of tourism from a local perspective, and thus to re-engage with some of tourism studies' most prominent dyads: global and local, colonization and decolonization. With a center-periphery model that has now lost its currency (Appadurai, 1986, 1996; Bhabha, 1994), it appears outdated to view globalization of tourism as the spreading of Western practices and imaginaries in the rest of the world. It rather appears to work as the constant rewriting of the rules by a broad array of actors who establish an active relation to this sphere of leisure that constitutes tourism.

Methodology

To answer these challenges, my methodology consists in the mutual observance of French colonial archives located both in France and in Vietnam, and field trips that allow me to conduct observations and semi-directive interviews with Vietnamese tourists in the main hill stations of the country: Sa Pa and Tam Đảo in the north, Bà Nà in the center, and Đà Lạt more to the south. Interviews were conducted in the interviewees' preferred language, mainly Vietnamese (with or without an interpreter), and otherwise English. The material for this chapter was gathered over a decade: I started collecting this material during my PhD geography thesis on domestic tourism in Vietnam, defended in 2009, and have been regularly updating it. My latest fieldwork in Bà Nà and Sa Pa was in 2016. While this period has witnessed a steady growth of domestic tourism in theVietnamese mountains, the long-term perspective allows for the consideration of the subtle evolutions of practice and place.

1 Taming the Mountains: Between the Wild and the Domestic

1.1 Mountains for the Kinh people: A paradoxical imaginary

As evidenced in Kinh folktales and toponyms, mountains are traditionally seen as dangerous spaces where mysterious and dramatic events occur. They are said to be occupied by magical creatures, both evil and good, hermits, fairies, or old and solitary outcasts. (Do Lam, 2007). In classical Kinh literature, homeland is fundamentally depicted as a village in an agricultural plain, an organized society with domesticated animals. This representation is in stark contrast with the threatening, wild or supernatural life of the mountains. Toponyms still account for such representations. In fact, the very expression designating mountains, *Nước Độc*, means both poisoned water and toxic land. This transcribes a history of diseases that plagued mountain regions, such as malaria. Locally, numerous toponyms mirror the mysteriousness and danger of mountains for the Kinh people. In Sa Pa for instance, the *Hàm Rồng* peak and eponymous garden (Figure 3.2) means the jaws of the dragon; in the center of Vietnam, the mountain that separates Huế and Đà Nẵng is called Đèo Hải Vân, the cloud pass.

Although they are seen as dangerous and mysterious spaces, mountains are nevertheless part of the Kinh dwelling. They are made beautiful through aesthetic processes, like *Hòn Non Bộ*, extremely codified miniature ornamental mountains: stones are harmoniously placed in a basin and bathed in water, on which are nested a pagoda, a house or a bamboo bridge, symbolizing a sporadic yet reassuring civilization. Fauna is also carefully picked: edible fish, a bird that symbolizes freedom, and so on. *Hòn Non Bộ* don't depict any specific landscape but rather display a tamed mountain made up of a selection of archetypal elements (Berque, 1995; Van Lit & Buller, 2001). This trend can also be seen in the handicraft of the imperial court – for instance the blue porcelain dishes from Huế. These aesthetics, inspired by Chinese Feng Shui, suggest a harmonious environment of water and mountains, subtly civilized with a pagoda, a fisherman or a farmer.

Both *Hòn Non Bộ* and imperial artifacts extend a past imaginary of mountains that is now popularized through touristic communication. Indeed, touring agencies make broad use of these imaginaries and may include foggy mountain landscapes, waterfalls, lakes and pagodas in their brochures.

1.2 Gardens and prayers: Making adventure familiar

Mountains can be seen as a complementary space for the Kinh people. According to Taoist teachings, the mountain is a kind of refuge. This explains the location of Buddhist meditation centers, which are preferentially at high altitude. Popular ones include Trúc Lâm, on the side of lac Tuyên Lâm, near Đà Lạt, or Tây Thiên, next to Tam Đảo. The mysteries surrounding certain peaks are also a source of cults and superstitions on longevity, prosperity or fertility. This also explains the presence of Vietnamese pagodas in the highlands. The two most famous ones are the Marble Mountain in the center of the country (Ngũ Hành Sơn) and the Perfume Pagoda, in the north (chùa Hương, Figure 3.1).

If tourism and pilgrimages are distinct, then sacred and profane practices are articulated in a complex way today in Vietnam, as in the rest of Asia (Gladstone, 2005; Guichard-Anguis, 2011). It is interesting to see that some skills are common to the two. For instance, fluvial or pedestrian itineraries used for pilgrimages are the same as those used for tourism. The same goes for guides and accommodation. In a way, pilgrims taught the way of the mountains to Kinh people. The Perfume Pagoda during the new lunar year is a space of prayer and recollection, but it is also a space used for walking, relaxing, sightseeing (with the cable car) and shopping. In Bà Nà, many domestic tourists visit the Great Buddha pagoda (Chùa Linh Ứng), or the Goddess of the Forest pagoda in Tam Đảo (*Chùa Thượng Ngàn*). The tourists pray to local divinities and bring offerings, not only so that their trip goes well but also to earn a favor or settle a problem. Such spiritual moments are not incompatible with more recreational activities following the pagoda's visit.

Figure 3.1 While mountains cover two-thirds of the country, Kinh tourism only occupies a handful of sites: Lunar New Year gathering in the Perfume Pagoda (photograph by E. Peyvel, 2006)

A second ancient practice reactivated by contemporary tourism is 'Du Xuân' ('going for spring'). It involves frequenting the mountain in early spring, a celebration period with the lunar New Year. During this period, tourists can admire peach, plum and apricot blossoms, which symbolize the promise of life and the good things announced by the New Year. The frail resistance of the fruit trees to the last assaults of winter is a symbol of tenacity and a metaphor used in Confucianism to express devotion and loyalty (Hữu Ngọc, 2004). Their red and yellow colors are also popular because they are symbols of happiness and success for Vietnamese people, as in other Asian countries that have similar practices (*hanami* in Japan, *beotkkot kugyeong* in Korea). Flowers are undoubtedly important aesthetic elements of this mountain landscape: they are wanted, valued, photographed, picked, and offered by Kinh tourists. This is why specific places have been built in the mountains to meet these representations and practices: gardens. A beautiful mountain is indeed a tamed mountain, and gardens contain this reassuring nature, ordered and aestheticized according to Kinh codes. They have selected flowers, *Hòn Non Bộ*, walkable paths, panoramas, etc., as shown in Figure 3.2, taken in *Hàm Rồng* garden in Sa Pa (in the north of the country). Looking at other mountain activities, such as hiking, it appears that Vietnamese tourists mostly ignore this practice (Peyvel, 2016). In Sa Pa, Tam Đảo, Bà Nà or Đà Lạt, while Northern Americans, Australians and Europeans use ethnic minorities' homestays along the hiking paths, the Kinh tourists rather focus on the hill stations, which are domesticated and tidy.

Gardens and praying are two mean by which the Kinh people sought to overcome the paradox of their presence in the mountains. They both encapsulate their inner motives: to confront the border of their conception of humanity, to venture in a space that was constructed as fundamentally wild only to domesticate and tame it. While the Kinh

Figure 3.2 Gardens, the orderly nature of Kinh mountain tourism (photograph by E. Peyvel, 2007)

imaginary of mountains is frightening, their touristic experiences are more exciting than actually dangerous. In that sense, the mountain has become a landscape (Figure 3.2), although this expansion is in fact limited to a few places: hills stations, lakes, gardens and temples.

2 Reclaiming the Colony: Between Oppression and Romance

Vietnamese hill stations were first built during the colonial period (1858-1954) for the settlers. Vietnamese people were in fact mostly excluded from these sites. However, despite this painful past, the stations were not only conserved after independence but also reoccupied. The paradox is twofold. Reclaiming these sites involves reclaiming the colonial thinking patterns, specifically the primitivism and exoticism/eroticism projected on ethnic minorities. Besides, hill stations are now associated with French romanticism rather than French colonialism. By stripping them of their historical political meaning, they can be a valid destination for a honeymoon, for instance.

2.1 From mimicry and exclusiveness in the colonial hill stations to an inclusive and innovative reclaiming in a postcolonial context

During the 20th century, mobilities in Vietnamese mountains took on another dimension, influenced by French colonization. Like other colonial empires (Spencer & Thomas, 1948), the French government decided to invest in tourism to maintain settler presence (Brocheux & Hemery, 2001). Tourism trips to the mountains were considered to contribute to the new settlers' mental and physical health, as they recreated a sense of 'home', even though they were far from Europe (Gaide, 1930; Sorre, 1943). Some places were built specifically with this purpose, notably seaside resorts and four hill stations: Sa Pa and Tam Đảo in the north, Bà Nà in the center, and Đà Lạt more to the south.

The French and Vietnamese archives on hill stations show the spatial model deployed in Vietnam is not new. Indeed, it is very similar to hill stations created at the same time in Europe. This is not due to a lack of imagination but, rather, reflects a deliberate intention to recreate France in French colonies (Demay, 2014; Jennings, 2011; Peyvel, 2009). The hill stations embodied the French representations of the mountain. Settlers could then recreate themselves by recreating their homeland.

It is possible to analyze the similarities between these colonial hill stations and the mainland on several scales. Firstly, regarding the location of these stations, tourists of that time found in the mountains the fresh and dry environment they were used to in France. Mountains allowed escape from the hot and humid cities. For example, Sapa is compared to the Alps for its fresh air. Regarding the stations' architecture, a French style was recreated: the Dalat railway station mimics the one of

Deauville; there are also Basque cottages and Breton houses, recalling the colonists' homeland. The similarities between the mainland and the colonial hill stations can also be analyzed in the local development plan, as illustrated in the map of Dalat (ASIE177, 1951), with the promontory along the central lake, the church, a sports center, a stadium and villas. In a broader sense, the whole mountain is set up for tourism, equipped with viewpoints and hiking trails, but also experimental farms to grow typically French vegetables and fruits (such as strawberries and artichokes), and also to raise cows to make butter, cream and cheese. Finally, the pace of these spaces made them typically French. They were frequented mainly during summer and weekends, Sunday mornings were dedicated to mass, and the rest of the week was animated by sporting events, walks, concerts, film screenings, chats and board games in the evening. The tourism development of Indochinese mountains was staged to comply with the aesthetic codes of the French settlers.

Colonial hill stations were both mimetic and exclusive in that they attempted to recreate an idealized France for the benefit of white people. In other words, at the heart of the colonial project regarding tourism is recreating Sameness to escape the Otherness of daily life in Indochina for the French settlers. While an indigenous elite was tolerated for profitability purposes, this concerned only a small fringe of people that was restricted to certain places, mainly hotels. In that sense, hill stations were conceived as extra-territorial entities, justifying sharp social and spatial closing. These stations also carried with them the ideology of their creators and are tangible manifestations of the triumphant modernity wanted by the colonial project. They were underpinned by the concept of Progress claimed by the urban planners who shaped them, such as Ernest Hébrard, the father of Dalat (Wright & Rabinow, 1982). From a Vietnamese perspective, there was a profound oddity in such a project that aimed to bring the comfort of the cities to the mountains.

Hill stations are also exemplary of the racial context of their time. As places set apart, they allowed escape from the otherness that Indochina exacerbated. For that purpose, Da Lat hill station's zoning scheme (1924 and 1932) established a double segregation: a functional one, that separated administrative, residential and commercial sectors; and a racial one, that allowed containment of Annamites. In that sense, zoning was a powerful tool to fulfill a racist plan.

Since independence, none of the colonial hill stations has disappeared: all still exist and are fully functional. Sa Pa and Đà Lạt have even experienced population growth, reaching respectively 9000 and 406,000 inhabitants. Most of them are strictly touristic sites, such as Tam Đảo, Sa Pa and Bà Nà, while Đà Lạt has managed to attract other types of activity, including a university. However, the fact that these stations are still functional should not mask the complex reclaiming processes they have gone through since independence. In terms of visitors, the number of Vietnamese is now far greater than foreign visitor numbers, while the French have almost disappeared. In recent years, Sa Pa and Tam Đảo have each hosted an average of 1 million tourists. Bà Nà, which was a declining station in the 1990s, received 2 million visitors in 2016. Hill stations have therefore gone from exclusive reproductions of colonial homeland to novel and inclusive sites frequented mainly by Vietnamese who deploy their own practices and imaginaries.

The reclaiming of places in the postcolonial context implies fundamentally hybrid practices. Certain places are now appreciated for the same qualities that the French first visited them for, such as the freshness of the summer days and the healthy benefits of mountain water. It is also interesting to note that certain practices are continued, for example hiking and wild camping, especially for young city dwellers who have the economic and cultural resources to develop distinctive practices of leisure (Peyvel, 2016). Simultaneously, Vietnamese tourists have also developed original practices made by borrowing and mixing bits and pieces, such as the syncretic trip, both spiritual and touristic, described earlier.

2.2 Primitivism as a colonial legacy: The Kinh tourists and the abiding exoticism of racialized populations

For Kinh people, taming the mountains in a recreational perspective does not concern only colonial places and practices, but also the people who live there. In a postcolonial context, sightseeing in the mountains for Kinh people means (re)claiming the place of the dominant ethnic group.

Other ethnic groups are less enthusiastic when it comes to mountain tourism and consider the Kinh trend with either incredulity or contempt. My field observations demonstrate that minorities are rarely photographed, contrary to western tourists whose lenses are particularly attracted by mountain natives, particularly women, children and the elderly. Besides, direct encounters are uncommon for domestic tourists, whereas they constitute a prime motivation for western visitors, notably those who practice trekking and home staying in Sa Pa (Peyvel, 2016). However, in Sa Pa and Đà Lạt, Kinh tourists frequently rent traditional clothes of various ethnic origins, not specifically regional. They enjoy being photographed with such attire. This could be interpreted as a step towards otherness, although this staging is usually as far as the inter-ethnic exchange goes. Persistent racial biases generate fear, misunderstanding and ignorance, even among the most educated layers of the population, as this interview excerpt indicates:

> Minorities are savage and innocent. They are not used to modern life. They know nothing about trade. We have to explain to them that working is not only slaughtering wild animals and chopping wood. (Hằng, 32 years old, teacher in Hà Nội)

The exercise of Kinh power involves reclaiming the codes of ethnic minorities' exoticization, such as referring to the 'savage' people, introduced by French settlers. This exoticism is fundamentally ambivalent (Said, 1978; Staszak, 2008) in that it consists of giving value to the authenticity, traditions and proximity with nature that these groups have, 'proof' of their primitivism, while building their otherness as a potential source of danger (especially towards national unity) and as a manifestation of their inferiority. The ethnology museums around the country are representative of this ambivalence, especially in Hà Nội and Buôn Ma Thuột in the central highlands. This exoticism serves to justify Kinh leadership in political and economic affairs, from the national to the local scale.

This exoticization of mountain peoples by the Kinh translates into some of their tourism practices. Previously noted links between exoticism and primitivism in postcolonial contexts (Coulouma, 2018; Fiskesjö, 2015; Hall & Tucker, 2004) are evident, for example, in the case of *Sinh Hoat lửa Trai*, which are fires around which Kinh tourists dance, sing and feast during their stay in the mountains. These celebrations are usually organized by hotels that dedicate a part of their outdoor space to this practice (Figure 3.3). This specific place allows the development of an unusual practice in a reassuring context for Kinh people. More than just a festive moment, these fire camps are particularly appreciated for what they depict of an imagined primitive life. The allegory of a wild life, and the more or less fantasized lifestyle of 'minority' ethnic groups, are mashed up in a kind of Elsewhere that Kinh tourists perform during those evening camp fires. The decorations of such accommodation (Figure 3.3), and the costumes that are used, are representative of this blend: Kinh tourists dress with clothes and objects of Bà Nà, Gia Rai and Ê De groups, such as gongs (classified by UNESCO in 2005). These *Sinh Hoat lửa Trai* are also interesting to study in the way that some heterosexual men live them to perform their masculinity: they drink a lot, sing loudly, dance shirtless and seduce or harass women. Everything happens as if their behavior was legitimized by a tourist practice that links the mountain and its inhabitants with unrestrained feral savagery.

Linking exoticism and eroticism, another practice stemming from exoticism sits on the border between tourism and prostitution. It results from the ambiguous attraction that some Kinh heterosexual men have for ethnic minority women. In colonial times, these women were both desired and vilified as wild animals. Even today, we can sometimes hear Kinh people comparing Hmong people and monkeys (*con khỉ*). In this postcolonial context, the exoticized and the exoticizers are Vietnamese: ethnic and gender criteria now feed this eroticism of domination. Because of the intersectionality in which these women are positioned, their gender and ethnicity are constructed as a source of otherness, justifying their inferiority by presenting them as wild women. This imaginary is the key to a prostitution economy, visible in Lào Cai, the

Figure 3.3 Postcolonial primitivism in Bà Nà (photograph by E. Peyvel, 2008)

provincial capital, where the highway and the train from Hà Nội for tourists willing to visit Sa Pa converge: notable is the presence of hotels, where rooms are rented by the hour, as well as karaoke and massage parlors, where prostitution also occurs. The city of Lào Cai is a double interface: at a national level, between Kinh and other ethnic groups of the northern mountains, and at a transnational level, taking advantage of the active Chinese border (Chan, 2013). In both cases, it is the exoticism of these women that is underpinning these markets: a Vietnamese for a Chinese, an ethnic minority woman for a Kinh. In Sa Pa, the love market – a historical meeting point for the local ethnic groups' youth – also exercises an erotic attraction (Michaud & Turner, 2006): for some Kinh, it is a place where sexual promiscuity can be easily obtained.

During the interviews I conducted, exoticism appeared in either ironic or implicit ways. For example, when I asked a group of six young Hanoian friends why they came to Sa Pa, Nam replied: 'To marry a H'Mong!' (*để lấy con gái H'Mong!*). This answer triggered heartfelt laughter by his comrades, as it is impossible to imagine a Kinh man marrying one of these women. Behind the laughter, there was also an underlying desire. Sometimes the aim of having intercourse was even more explicit with the expression of *bắt vợ* (to bang her). These Kinh heterosexual men allow themselves this register of words and sexual practices even more so as tourists far from their daily social environment, where social regulation is stronger. Several studies on the subject (Wah, 2013; Zhixiong, 2003) show that having sex with a woman of 'minority' ethnic group may be rewarding, flattering the customer's virility. Some of them like to be defined as hunters, and develop a conquering warrior discourse, while others suggest they only respond to their wild and insatiable assaults. In both cases, women are compared to impetuous animals. It is the colonial stereotype associating indigeneity with an unbridled sexuality that is revived.

Vietnam is not alone when it comes to the continuation of colonial schemas and colonial eroticism. In his study of *nagsu katun lem la bat* (cartoons sold in Thailand), Baffie (1989) highlighted the resemblance between the way women of minority groups are exoticized in some explicitly erotic pages and the exoticism depicted in National Geographic, a magazine rooted in colonialism, read by American soldiers during the Vietnam war and accustomed to publishing pictures of nude women under the cover of ethnography.

2.3 French romanticism revisited: Vietnam's membership within global consumerism

If French colonization led to tourist globalization, this process is continued today with the circulation of global leisure models. In this case, France is often associated with love and romance, making it an ideal honeymoon destination. This is a relatively new kind of trip for Vietnamese couples, and domestic developments have emerged to meet these expectations, especially in Đà Lạt and Bà Nà, recreating a small France, favorable to romantic honeymoons (*Tuần trăng mật*). For the middle-class Vietnamese tourists who cannot afford the trip to France, these sites recreate a well-known French otherness, based on the circulation of globalized models of leisure.

In Đà Lạt, the Eiffel Tower-shaped radio antenna has earned it the name of *Petit Paris*. Tourists can ride in a carriage around the Hồ Xuân Hương lake (the name of a poetess who wrote suggestive poems in the 18th century, like *Quả mít*, The Jackfruit). Outside the city, tourist sites frequented for their romanticism are mainly lakes and waterfalls, serving as the backdrop for wedding pictures. There is also 'La vallée de

Figure 3.4 The colony as a romantic setting in Bà Nà (photograph by E. Peyvel, 2016)

l'amour' (valley of love), a garden dedicated to these practices. Đà Lạt is not only romantic for Vietnamese tourists but also for Japanese, for whom the city was popularized in the Fumiko Hayashi novel, *Ukigumo* (floating clouds). In both cases, the misty landscapes are associated with a romantic atmosphere.

While Đà Lạt is the hill station that boasts the most colonial heritage, it is not the only stage for romantic ventures: artificial and fake places were built recently for this purpose. For example, in Bà Nà, a French Village (*Làng Pháp*) was built. Designed by an American firm, its artistic direction was entrusted to a South African company, and the architecture was supervised by a French company. An entire French village was recreated, where medieval, Renaissance and colonial times are mashed up in a romantic way (Figure 3.4). There are hotels, restaurants, cafés, shops and a church. A 'jardin de l'amour' was created, full of pink hearts, echoing that of Đà Lạt, where we find again this conception of a tamed mountain nature. Those gardens blend the spirituality of the pagodas and the Sinicized aesthetic of Hòn Non Bộ, discussed previously.

Conclusion

Historically, mountains are the border of the Vietnamese nation and were only recently included in the national story. Their mysteriousness and danger made them unattractive for the Kinh people, who only saw their spiritual potential, leaving the other 53 minorities to occupy these regions. The recent rise in domestic mountain tourism composed and controlled by the Kinh is therefore paradoxical in that it entails spatial, socio-ethnic and representational transgressions.

The surge in domestic tourism since the Đổi Mới, increasing from 9.6 million in 1998 to 85 million tourists in 2019, also multiplied the destinations available in the country. However, while mountain tourism is usually a quest for space, Vietnamese domestic tourism is highly concentrated in only a few sites, such as pagodas, temples, and hill stations in which the wilderness is tamed and ordered through landscape construction, gardens and lakes.

The growth of Kinh mountain tourism is also paradoxical in that it reclaims colonial sites, practices and imaginaries. Kinh people have developed an inclusive and innovative way to reclaim tourism in mountains from an exclusive and mimetic model developed by the French during colonization. However, this reclaiming does not break from the colonial way of thought. On the contrary, the paradox may be elucidated when considering the power relationship between Kinh people and dominated mountain ethnic groups who are exoticized through tourism. Besides, the authenticity of French heritage has not always been the subject of a protection policy in these hill stations. The postcolonial reclaiming goes hand in hand with selective memory of the colonial past and is therefore made both of silences and reinterpretations. This explains why hill stations built by settlers now serve as romantic scenery despite the painful memories they may uphold. In that sense, Kinh mountain tourism tackles the key paradoxical dyads of tourism studies: colonial and postcolonial, centering and decentering, global and local, expansion and limitation, wild and domestic, adventure and banalization.

References

Appadurai, A. (1986) Theory in anthropology: Center and periphery. *Comparative Studies in Society and History* 28 (2), 356–361.
Appadurai, A. (1996) *Modernity at Large: Cultural Dimensions of Globalization*. Minneapolis, MN: University of Minnesota Press.
Baffie J. (1989) Highlanders as portrayed in Thai penny-horribles. In J. McKinnon and B. Vienne (eds) *Hill Tribes Today: Problems in Change* (pp. 393–407). Bangkok: White Lotus & ORSTOM.
Berque, A. (1995) *Les raisons du paysage, de la Chine antique aux environnements de synthèse*. Paris: Hazan.
Bhabha, H. (1994) *The Location of Culture*. New York, NY: Routledge.
Brocheux, P. and Hémery, P. (2001) *Indochine, la colonisation ambigüe (1858–1954)*. Paris: La Découverte.
Bruneau, M., Taillard, C., Antheaume, B. and Bonnemaison, J. (1995) *Asie du Sud-Est: Océanie*. Paris & Montpellier: Belin/GIP Reclus.
Chan, W.C. (2013) *Vietnamese-Chinese Relationships at the Borderlands: Trade, Tourism and Cultural Politics*. New York: Routledge.

Coulouma, S. (2018) Une ethno-histoire des Wa-Paraok de Wengding (Yunnan, Chine): Pratiques, représentations et espace social face au tourisme. PhD thesis, Université d'Aix-Marseille.
Demay, A. (2014) *Tourism and Colonization in Indochina (1898–1939)*. Newcastle: Cambridge Scholars Publishing.
Do Lam, C.L. (2007) *Contes du Viêt Nam : Enfance et tradition orale*. Paris: L'Harmattan.
Fiskesjö, M. (2015) Wa grotesque: Headhunting theme parks and the Chinese nostalgia for primitive contemporaries. *Journal of Anthropology* 80 (4), 497–523.
Gaide (1930) *Les stations climatiques en Indochine*. Hanoi: Imprimerie d'Extrême-Orient.
General Statistics Office of Vietnam (2016) *Vietnam Population and Housing Census*. Hanoi: Nhà xuất bản thống kê (Statistical publishing house).
Ghimire, K. (2001) *The Native Tourist*. London: Earthscan.
Gladstone, D.L. (2005) *From Pilgrimage to Package Tour: Travel and Tourism in the Third World*. New York, NY: Routledge.
Graburn, N. (1983a) *To Pray, Pay and Play: The Cultural Structure of Japanese Domestic Tourism*. Aix-en-Provence: Centre des hautes études touristiques.
Graburn, N. (1983b) Anthropology of tourism. *Annals of Tourism Research* 10 (9), 9–34.
Guichard-Anguis, S. (2011) Walking through world heritage forest in Japan: Kumano pilgrimage. *Journal of Heritage Tourism* 6 (4), 285–295.
Hall, C.M. and Tucker, H. (2004) *Tourism and Postcolonialism: Contested Discourses, Identities and Representations*. London & New York: Routledge.
Hitchcock, M., King, V.T. and Parnwell, M.J.G. (eds) (1993) *Tourism in South-East Asia*. London & New York: Routledge.
Hữu Ngọc (2004) *Wandering through Vietnamese Culture*. Hà Nội: Thế giới publishers.
Jennings, E. (2011) *Imperial Heights: Dalat and the Making and Undoing of French Indochina*. Berkeley, CA: California University Press.
Koninck de, R. (2016) *L'Asie du Sud-Est*. Paris: Armand Colin.
Michaud, J. and Turner, S. (2006) Contending visions of Sa Pa: A hill-station in Viet Nam. *Annals of Tourism Research* 33 (3), 785–808.
Nash, D. (1981) Tourism as an anthropological subject. *Current Anthropology* 22 (5), 461–481.
Papin, P. (2003) *Vietnam, parcours d'une nation*. Paris: Belin.
Peyvel, E. (2009) Tourisme et colonialisme au Vietnam. In C. Zytnicki and H. Hazdaghli (eds) *Le tourisme dans l'empire français: Politiques, pratiques et imaginaires (XIXe-XXe siècles)* (pp. 133–143). Paris: Société française d'histoire d'outre-mer.
Peyvel, E. (2016) *L'invitation au voyage:* Géographie postcoloniale du tourisme au Vietnam. Lyon: ENS éditions.
ASIE177 (1951) *Plan de Dalat et environs au 1/10 000e*. Aix-en-Provence: Archives Nationales d'Outre-mer (ANOM).
Said, E. (1978) *Orientalism*. New York, NY: Pantheon Books.
Scott, J.C. (2009) *The Art of Not Being Governed: An Anarchist History of Upland Southeast Asia*. New Haven & London: Yale University Press.
Singh, S. (2009) *Domestic Tourism in Asia: Diversity and Divergence*. London: Earthscan.
Sorre, M. (1943) *Les fondements biologiques de la géographie humaine*. Paris: Armand Colin.
Spencer, J.E. and Thomas, W.L. (1948) The hill station and summer resorts of the Orient. *Geographical Review* 38 (4), 637–651.
Staszak, J.F. (2008) Qu'est-ce que l'exotisme? *Le Globe* 148, 3–30.
Teo, P., Chang, T.C. and Ho, K.C. (2001) *Interconnected Worlds: Tourism in Southeast Asia*. Oxford: Pergamon.
Tucker, H. and Akama, J. (2009) Tourism as postcolonialism. In T. Jamal and M. Robinson (eds) *The Sage Handbook of Tourism Studies* (pp. 504–520). London: Sage.
Urry, J. (2011) *The Tourist Gaze 3.0*. London & Los Angeles: Sage.
Van Lit, P. and Buller, L. (2001) *Mountains in the Sea: The Vietnamese Mountains Miniature Landscape Art of Hon Non Bo*. Portland, ON: Timber Press.

Vietnam National Administration of Tourism: http://www.vietnamtourism.gov.vn/english/
Wah, C.H. (2013) *Vietnamese-Chinese Relationships at the Borderlands: Trade, Tourism and Cultural Politics*. London & New York: Routledge.
Winter, T., Teo, P. and Chang, T.C. (2009) *Asia on Tour: Exploring the Rise of Asian Tourism*. London & New York: Routledge.
Wright, G. and Rabinow, P. (1982) Savoirs et pouvoirs dans l'urbanisme colonial d'Ernest Hébrard. *Cahiers de la recherche architecturale* 9, 27–44.
Zhixiong, H. (2003) Migration and the sex industry in the Hekou-Lao Cai border region between Yunnan and Vietnam. In D. Muhadjir, A.M. Wattie and S.E. Yuarsi (eds) *Living on the Edges: Cross-border Mobility and Sexual Exploitation in the Greater Southeast Asia Sub-region* (pp. 1–44). Yogyakarta: Gadjah Mada University Press.

4 The 'Logical Paradox' of Preservation via Change: The Touristic Potential of Malaysia's Catholic Mission Schools

Keith Kay Hin Tan and Paolo Mura

Introduction

Beginning with the Portuguese conquest of Malacca in 1511, the Catholic church has been active in setting up schools in what is now Malaysia. Originally the purview of Portuguese Jesuit priests, Catholic schools expanded dramatically during the British colonial period in the late 19th and early 20th centuries via the influence of French and Irish missionary teachers. Statistics from the Catholic Research Centre, Kuala Lumpur indicate that at the peak of their relative influence, these 'mission' schools graduated almost half of all English-educated secondary students in then Malaya in the 1950s (Tan, 2011). Enjoying a high degree of academic and administrative autonomy during the colonial period, mission school privileges were gradually eroded by independent Malaysia's ministry of education after wide-ranging changes to the country's education system were introduced in 1970 (Chew, 2000). Together with a dramatic fall in Catholic religious vocations worldwide following the completion of the Second Vatican Council in 1965, significant changes occurred in the character of the mission schools. This caused Malaysia's Catholic church to dissolve its 'Guild of Assisted Catholic Schools' in 1989, when it acknowledged that it could no longer adequately control the appointment of school principals (Pappu, 1996). In turn, this effectively transferred administrative control of more than a hundred large schools from the Catholic church to the Malaysian government upon the retirement of their missionary principals from the 1980s onwards, so that by 2008 all these previously autonomous schools had been absorbed into the mainstream of Malaysia's public education system (Ward & Miraflor, 2009).

Because of the fading of the missionary legacy, there are now conflicting opinions from older alumni and the current, government-appointed stakeholders about the evolving heritage of these historically Catholic schools, the oldest of which occupy prominent sites and architecturally significant buildings. This is, in part, because multiple narratives exist concerning the identity of these places. Although heritage status has been conferred on the oldest schools by the Malaysian government, most do not attract touristic interest until they cease to accept students. In the handful of cases where this has happened, the schools' tangible architectural heritage has then been successfully re-interpreted for tourism. Indeed, although digital applications such as Tripadvisor and Wikitravel are extremely useful in increasing the popularity of existing heritage tourism 'products', the re-invention of existing, non-touristic yet culturally significant sites as tourist attractions is under-researched in contemporary tourism studies (Tan, 2017). This chapter argues that these places are worth studying in general as the 'starting point' of tourism – acting as a kind of 'potential energy', where 'tourist-friendly' narratives are waiting to be written.

This chapter explains how a tourism-led change of function can be a useful catalyst for the 'preservation' of a historical site's heritage, which would paradoxically be threatened by the 'continuity of use' so often celebrated by tourism scholars and heritage bodies (International Council on Monuments and Sites (ICOMOS), 1994). The 'logical paradox' of this argument lies in the fact that 'change' and 'preservation' are opposites, yet to preserve a historical site as a heritage tourist attraction, some degree of change is often necessary.

This paradox is especially the case for Malaysia's mission schools, which are among the oldest built structures in the country. The chapel of a former mission school in Malacca is in fact the oldest known European building in Southeast Asia. Now known as St Paul's Church, its 16th-century ruins lie at the heart of Malacca's United Nations Educational, Scientific and Cultural Organization (UNESCO) World Heritage and tourism district (Figure 4.1). Its academic heritage has, however, been almost forgotten due to multiple changes of use since the founding of what was originally the Jesuit 'College of St Paul' in 1548 (Tan, 2015). Across major towns and cities on Malaysia's west coast, many mission schools have likewise been granted 'heritage' status without any real plan to safeguard their futures, beyond a commitment not to demolish their original buildings.

To some extent, this can be attributed to the fact that what counts as heritage (and especially, what counts as 'cultural' heritage), has generally been decided by 'experts' (Throsby, 2001). These 'professionals' include, but are not limited to, art historians, archeologists, museum curators,

Figure 4.1 Ruins of St Paul's Church, Malacca, earlier part of a Jesuit school started in 1548 (photograph by main author)

conservationists, urban planners and architects. It is a list that largely ignores the opinions, desires and needs of 'ordinary people' – even ordinary people with direct and long-lasting connections to the sites in question. In the case of Malaysia's mission schools, this has resulted in a loss of intangible heritage which, as will be described later in this chapter, is evident to all interviewed research participants who graduated from such places before 1989.

The heritage mission schools (or remnants) which this chapter derived the bulk of its data from, with the exception of the Cameron Highlands convent, all have 'listed' status preventing demolition, granted at either local, state of federal government level, or are located in areas under overall development control.

As they are amongst the oldest extant schools in Malaysia, the elevation of these buildings to national heritage status in the last few decades points to a research gap about what 'mission school heritage' actually represents, especially with regards to tourism. By examining the oral histories of the schools from the point of view of the alumni and former teachers who are most interested in safeguarding the schools' identity, this chapter argues that paradoxically, a change of function can become a necessity for heritage preservation.

Tourism in Post/Neo-colonial Societies: 'Accepted' Identities and Mission School Heritage in Malaysia

One of the arguments often reiterated in the tourism literature of the last 20 years (Hall & Tucker, 2004; Hollinshead, 2004; Mura & Wijesinghe, 2019; Wijesinghe *et al.*, 2019) concerns the role of tourism in reproducing, reshaping or questioning colonial/postcolonial/neocolonial hegemonic power relations. Indeed, it has been claimed that tourist spaces (especially those in non-Western territories formerly colonized by European powers) are often re-constructed and re-presented according to postcolonial/neocolonial power structures, which tend to reinforce stereotypical images of local communities as 'primitive', 'pre-modern' and 'pre-industrial' (Hall & Tucker, 2004). This process, in turn, contributes to the re-shaping of individual, local and national identities, which must then be renegotiated and/or reinvented according to tourism developmental and promotional strategies (Palmer, 1999).

This is important, because the myths underpinning tourism promotional material are crucial to encouraging tourist consumption as they often fulfil Western fantasies and desires concerning the 'romantic' and 'primitive' *other* (Hall & Tucker, 2004). Tourism may therefore act as an agent reiterating postcolonial/neocolonial discourses and stereotypical subordinated representations of gender, class, ethnicity and race. This is particularly true within the context of heritage tourism attractions and contexts (Ashworth & Tunbridge, 1996), which in former colonies often have colonial associations (Henderson, 2004).

It is important to emphasize that heritage tourism experiences and products are always informed by selected tourist narratives and/or cultural tourist objects, but the choice of what does or does not constitute heritage for tourist consumption often involves politically charged decision-making. Inevitably, therefore, some aspects/objects of specific local cultures eventually occupy privileged/preferred positions vis-à-vis others (Ashworth & Tunbridge, 1996). Moreover, political choices also underpin the interpretation of tourist heritage sites/tourist objects (Henderson, 2004). In this respect, while certain narratives, photos and videos assume a dominant role in framing and directing the tourist gaze, others are silenced or 'forgotten'. This process does not always occur unquestioningly or unproblematically: rather, it often leads to conflicts among different parties/perspectives. As Henderson (2004: 113) pointed out: 'postcolonial societies have to confront recent experiences of occupation and external control with the potential for conflicts over the narratives to be communicated about the past to contemporary audiences, both domestic and international'.

Although in postcolonial realities tourism (specifically heritage tourism) represents only one of various social phenomena reproducing post/neocolonial hegemonies (Hall & Tucker, 2004), its political power should not be underestimated. It is within this conceptual frame, therefore, that we can examine how previously colonized territories, such as Malaysia, are 'produced' and promoted as heritage tourist destinations. Indeed, Desforges (2000) argued that identity issues lie at the heart of our desire for greater tourism travel, while Palmer (1999) suggested that heritage-rich, community-based sites help domestic tourists especially answer questions about their own origins. Where Malaysia's mission schools are concerned, however, these arguments must be read against the fact that national governments often place a higher value on some histories and cultures whilst devaluing others (Hall & Lew, 2009), with 'preferred' histories reflecting policy decisions to establish and promote an 'accepted' national identity (Kuipers & Ashworth, 2001).

The position of Malaysia's mission schools within the country's 'accepted' postcolonial identity is particularly unclear. Butler *et al.* (2012) identified the presence of a strong, ethnic and religious identity connection between religious heritage sites and their adherents in Malaysia, while there is a marginal indifference to non-religious sites, especially those designated as 'colonial' in nature. As educational buildings, mission schools are not, strictly speaking, religious sites, despite being host to Catholic chapels. However, their status as 'church properties', their period of construction and use of 'Western' architecture has often led to their labelling as both 'alien' and 'colonial', especially amongst non-Christian groups. This has led to more right-wing voices calling for the buildings to be re-named, altered or in extreme cases partially demolished, as was once suggested by two Malaysian government members of parliament during budget policy debates in October 2007 (Lim, 2007).

By contrast, heritage conservation movements in Malaysia have in recent years emphasized the importance of preserving and promoting (also for tourist consumption) colonial heritage, including mission schools, as a vehicle to increase social cohesion among the different ethnicities constituting Malaysia's diverse social fabric in order to foster national unity (Henderson, 2004). Overall, the different positions and views concerning the use, re-use or non-use of mission schools in Malaysia (both for tourist or non-tourist consumption) reflect the ambivalences and paradoxes underpinning the meanings and interests (political, economic and sociocultural) of colonial heritage.

Narratives and Participants

The research participants of this study, 15 in total, were interviewed individually from early 2015 to mid-2016. They were selected for the richness of their experience of the mission schools as

Table 4.1 Research participants (all names are pseudonyms)

Participant's pseudonym	Age	Ethnicity	Religion	Connection/background
Brother Peter	85	Irish	Catholic	Religious brother
Linda	70	Chinese	Catholic	Retired teacher
Francesca	64	Indian	Catholic	Former nun
Indra	65	Indian	Hindu	School alumnus
Lai Mei	65	Chinese	Taoist	School alumnus
Emma	76	Indian	Catholic	School alumnus
Sister Dorothy	75	Chinese	Catholic	Religious nun
Robert	60	Chinese	Protestant	School alumnus
Nathan	75	Indian	Catholic	School alumnus
Dr Lawrence	54	Chinese	Catholic	Former religious brother
Faisal	56	Malay	Muslim	Chairman, Board of Governors
Dr Aishah	54	Malay	Muslim	School alumnus
William	46	Chinese	Protestant	School alumnus
Sufina	58	Malay	Muslim	Retired teacher
Razak	56	Malay	Muslim	School alumnus

well as their enthusiasm to share opinions, as determined via snowball sampling starting with one of the last surviving foreign missionaries in Malaysia. All residing in the Kuala Lumpur area, they reflected Malaysia's major ethnic groups and both genders, ensuring the necessary breadth, depth and richness of experience to be worthy of the topic (see Table 4.1). The enthusiasm of relatively old participants to be interviewed further contributed to the significance of the data, given their previous administrative seniority in, and vast knowledge of, the majority of the heritage schools in question.

Focus of the Interviews

The focus of the interviews was to extract individual stories. By interviewing 15 participants with similar experiences, it was also possible to identify collective endeavours and to recognize individual and group contributions to the changing of culture and society which are vitally important in the critical discourse analysis (CDA) technique that was used to analyse the transcripts. Comparing, coding and identifying themes from these diverse data provided important validity checks, which lent further strength to the study.

The topics covered touched on religious as well as secular identity, where the former especially has changed dramatically over the last 50 years. Whereas much tourism research has focused on 'the tourist' as the pivotal centre of activity, this study looked at the 'source' issues of identity and heritage without which cultural tourism as an activity would not exist. The people interviewed were therefore not tourists but, rather, stakeholders of the schools who nevertheless had strong opinions about the tourism potential of the sites. The question of identity was treated as significant yet fluid, changeable over the lifetime of an individual, whereas tangible heritage was treated as permanent yet evolutionary, changeable over the existence of a historical site.

The interviews differed, therefore, from the 'life history' interviews common to many narrative-based studies. There was no attempt to capture the entire life story of the research participants. The focus of the oral history interviews was instead to gather testimony on the evolving heritage of Malaysia's mission schools from the point of view of alumni and former teachers, missionary and lay people alike, with the individual interviews conducted in a semi-structured fashion. In order to establish familiarity between participants and the researcher, the 'structured' portion of these interviews focused on identity-based questions such as:

- How has your experience of schooling affected your sense of personal identity?
- How did your schooling affect your career choice?
- What was the most satisfying (and unsatisfying) memory about your school?

After identifying how an interviewees' identity had been formed by his/her mission school experience, the next tranche of 'structured' questions shifted the focus from the individual to the schools themselves, via questions such as:

- What is the most unique/the most important characteristic of your/the mission school(s)?
- How important are the school buildings to their character, identity and heritage?
- How important was the foreign connection to the mission schools, and why?

Later, the focus shifted to 'value judgement' questions, to discover how participants viewed mission school heritage and its related constructs. For example:

- How would you describe mission school heritage?
- How have the changes to mission school character since the 1980s made you feel?
- Do you see mission school heritage as something that should be preserved, or something that should evolve?

The final tranche of questions focused on the future and/or participants' suggestions:

- What is your opinion regarding adaptively re-using/privatizing the mission schools?
- Can the mission schools have a valid, post-Catholic identity?
- What is your opinion about opening up the mission schools for tourism?

This brief interview guide for structured questions concentrated on acquiring descriptive, knowledge-based responses as well as 'opinion-based' or 'value' judgements from the participants. The 'unstructured' questions forming part of each interview were, however, just as important as the 'structured' component. Their nature depended on the length, depth and emphasis given by respondents to the initial set of structured questions. Responses to unstructured questions also changed the sequence of the structured ones. The goal was to maintain, at all times, the comfort level and the 'talkativeness' of the interviewees and to set the stage for a meaningful discussion about the role of tourism in maintaining mission school identity as defined by the participants themselves.

At the same time, digital technology has transformed the way history is being recorded, preserved and shared (Perks & Thomson, 2006). All live interviews were thus audio-recorded, before being manually transcribed into Word documents. This was important in order to retain the structure of individual narratives as a whole, in order to make meaningful later comparisons between information from different research participants. The longest interview lasted for six hours over two days, whilst the shortest interview lasted just under two hours.

Critical Discourse Analysis

Once the interviews began, it became rapidly evident that the participants' emotional attachment to their old schools was particularly strong, in many cases made more so by the passage of time. Additionally, it was also evident that participants shared a widely held perception that their opinions were largely ignored by Malaysia's ministry of education, leading to further negative emotional reactions. These contributed to the decision to use critical discourse analysis (CDA) to analyse the interview transcripts. Meyer (2001) premised that CDA must not be taken as a single method, but rather a data analysis approach requiring a number of selections to be made at different levels, with CDA scholars often playing 'an advocatory role for groups who suffer from social discrimination' (Meyer, 2001: 15). In tandem with the multiple approaches of CDA, there also exists work done by multiple researchers who have all contributed to the continuing development of critical discourse analysis.

Among the key researchers in the field of CDA, Fairclough stands out as one of the most influential (Amerian & Esmaili, 2015). Fairclough emphasizes 'social conflict ... and tries to detect its linguistic manifestations and discourses, in particular elements of dominance, difference and resistance' (Meyer, 2001: 22). In Fairclough's view, CDA is the 'analysis of the dialectical relationships between semiosis [any form of activity, conduct or process that involves signs, including language] with other elements of social practices' (Fairclough, 2001a: 122). He called this particular method of CDA a 'Critical Language Study'. Together with Chouliaraki, he premised that CDA creates '... awareness of what is, how it has come to be, and what it might become' before further stating that CDA '... systematically charts relations of transformation between the symbolic and non-symbolic, between discourse and the non-discursive' (Chouliaraki & Fairclough, 1999: 113).

Fairclough further premised in 2001 that spoken or written texts are the 'product' of a 'process', with the latter having two dimensions – 'production' and 'interpretation'. Texts end up as both the 'product' of the 'process of production' as well as the 'resource' of the 'process of interpretation' (Fairclough, 2001a: 21). The fact that most CDA scholars also view language as a means of social *construction* introduces a third, contextual element into the approach, resulting in what has come to be known as Fairclough's 'Three Dimensional CDA Model' (Fairclough, 2001b), an interpretation of which is shown in Figure 4.2:

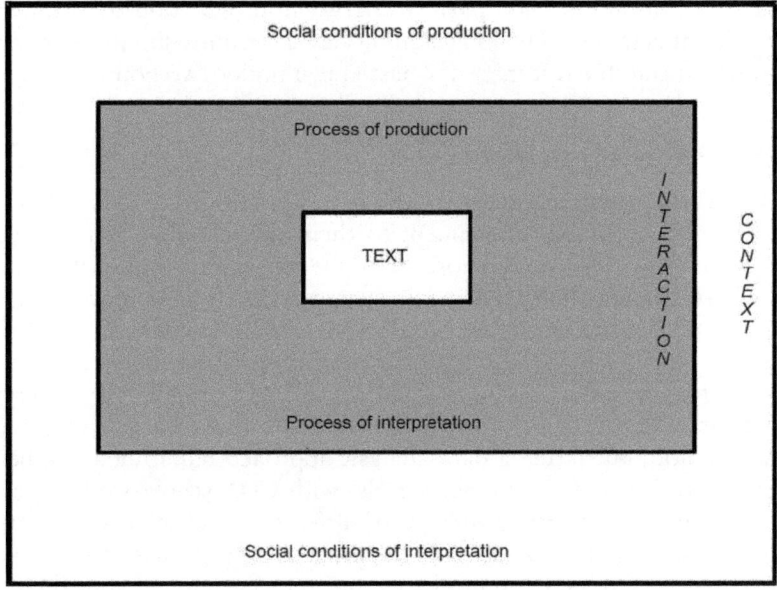

Figure 4.2 Fairclough's 'Three-dimensional CDA model' (as interpreted by main author)

An interpretation of Fairclough's approach gives rise to three 'analytical focal points'. These are, organized from the smallest to the largest squares in Figure 4.2:

(1) The text itself (whether written or spoken).
(2) The 'discourse practice' (i.e. the process of production and interpretation).
(3) The 'socio-cultural practice' (i.e. social and cultural structures surrounding communicative events).

This approach places an emphasis on concealed relationships between the three focal points, and, by extension therefore, to the links between language, power and ideology. This is especially relevant to a study about the construction and discursive re-construction of Malaysia's mission school heritage, because of the various changes in socio-political discourse from the colonial period, when the schools were founded, to the present day.

Mission School Narratives

Across the board, the participants held either positive or very positive opinions about the mission schools before they were subject to government control, with attitudes generated either from distinctive personal experiences or by a real or perceived contrast between how 'good things were' in the past to how 'bad things are' in the present; and this was true regardless of their religious background. Even where critical thoughts were present, they were measured against the numerous corresponding 'positives' and dismissed as relatively unimportant by the participants themselves.

The Importance of the English Language and the Foreign Connection

It was clear that many participants equated a good standard of English to not only a well-rounded education but also to a set of 'values' that many suggested were universal:

> I think when the Missionaries came, they gave a whole new level to the standard of English spoken ... it's unifying for the races. Malay, Indian, Chinese, and I don't have to say 'I speak more Chinese' ... 'I speak better Malay' ... but we all speak good English. It draws us away from our cultural origins ... (Lai Mei)

All participants were fluent and comfortable with English, indeed using it as their 'first language', suggesting an appropriation of English as something global yet also local, a language that some Malaysians regard as their own, despite it having been largely introduced by European missionaries. Apart from Dr Lawrence, all locals in fact credited the foreign connection for their positive 'state of mind' regarding the schools. This included various references to colonialism, both oblique and direct: 'I think the foreign mission set the standard against which the local schools had to benchmark ... that was very good ... at that time, everything European, we "looked up in awe"' (Indra).

The only Irish-born missionary to be interviewed explained that the traditional Catholic/Protestant divide in the British Isles mattered little during Malaysia's colonial era: 'I think the Brits were looking for people who could provide education to help their colonial rule, and who provided it they weren't too worried' (Brother Peter). This theme of cross-cultural co-existence in the mission schools would recur throughout many interviews, and was identified as a strength that, if exploited, could render the sites attractive to tourists from across the ethnic and religious spectrum.

Adopting Values, Adapting Cultures

Like the English language, another 'import' that was later localized was the missionary work culture, also attributed to their foreign background: 'Europeans are known to be very systematic in their way, very strict ... I think the Europeans brought ... a good work culture into the convents' (Sufina). A sense of pride was detected in many replies, further evidenced by a shift in tone, emphasis and body language during the interviews. Here, Muslim participants also made reference to how the schools' influence stretched beyond the classroom into other realms of public life in Malaysia compared with how government schools are being run today: 'The Missionary Schools ... the culture was different ... we were more broad-minded' (Sufina). The suggestion that the past carried a broader, more sensitive or better set of values was evident in all the respondents' answers, and also re-surfaced several times over. The idea that Christian values in general, and the mission schools in particular, had indirectly influenced later aspects of Malaysia's social development was further expanded into a discussion on symbolism and form.

Symbolism and Form

One of the participants suggested that people are naturally attracted to form: 'I believe that we human beings, though how much logic we like to apply in terms of function, we tend to be attracted by form' (Razak). Paradoxically, the interviews identified an inverse relationship between the depth of a participant's involvement with the schools and his or her attachment to their symbols. 'Structure' and 'visibility', or a lack thereof, were indeed recurring themes in the narrative stories. Perhaps surprisingly, non-Catholics were in general more supportive of and nostalgic about the mission schools' disappeared 'commanding presence' compared to Catholics, with devout Catholics amongst the least attached to the 'power structures' of the 'old ways'. This suggests an 'outsiders' attraction' to the mission school narrative. It also suggests a strong, secular appeal of the campuses and a wide interpretation of what they once stood for.

Mission School Heritage

The 'old ways' were greatly emphasized when participants were asked to describe 'mission school heritage' from their point of view: 'Heritage is something in the past … which still impacts on the present …' (Nathan). The removal of the 'missionary ethos' that had previously differentiated the schools from the rest of Malaysia's public school network was therefore almost universally condemned, as was the perceived suppression of links between the schools and the Catholic church and the introduction, in some cases, of a more Islamic ethos in keeping with the country's majority faith: 'It's our colonial roots … people want to get rid of these colonial roots, and "dis-root" us … for goodness sake!' (Dr Aishah).

Almost uniformly, participants described 'mission school heritage' either in the past tense or in contrast to the present. The same heritage was perceived as setting themselves 'apart' or even 'above' society at large. The perception that this heritage was disappearing served to place even greater value on it. This 'rarity premium' extended through both the schools' 'tangible' and 'intangible' heritage, with strongly perceived links to the schools' architecture and touristic potential, since even disinterested government control was seen as unable to erase the 'story-telling' ability of the 'authentic' school buildings and their potential to communicate symbolism and form to a new generation of otherwise unwitting visitors.

It was clear that all participants felt they had a stake in mission school heritage. The unifying effect of this on people of different ethnic and religious backgrounds was evident despite the occasional difference in interpretation or emphasis, with the schools' tangible heritage offering the most cohesive point of agreement:

> Form has always been a human being's natural mode … Important … reminds you of the past, of the history, how all this has come about … all the work that has been put in … we should not be doing things [if] we don't have the intention of being here for the long run. (Faisal)

Comments like this connected the tangible and intangible heritage of the schools by showing how the existence of one without the other diluted the meaning of both. The importance of maintaining at least one of these threads intact was expanded on by other participants as well:

> ... our memories are attached to the architecture. So if you remove that, you remove the legacy ... this is our heritage. Now people say colonial heritage is not really heritage, as if these are political interventions ... there are people who want to destroy the heritage, because heritage is contention ... like identifying oneself. (Dr Aishah)

The symbolism and form of the school buildings, even more than the vanished intangible heritage, represented an all-important *continuity* for the participants. Indeed, there was clearly a fear that any loss of the heritage buildings especially would lead to a 're-packaging' of history that would erase mission school identity in the same way the schools' character has already been diluted to the point of irrelevance. The participants identified tourism as one of the few industries that had the potential to save the old school buildings by 'monetizing' their history in a way that can convince all stakeholders about the value of preserving the past.

Socio-political Impact of Government Policy

Because the participants' fears for the future of the schools under government control were serious, they were asked to address the idea of 'evolution' versus 'preservation' of mission school heritage, to determine if a changed identity could still be 'valid'. Their responses showed an acceptance of evolution as an inevitable by-product of modernity as long as the 'core values' of the schools survived as guiding principles: 'Certainly it [mission school heritage] should be preserved, it's part of the history, the culture and the development of education in the country, but you also have to adapt to different times ...' (Brother Peter).

The responses again highlighted continuity as a core desire, that any evolution of mission school heritage should not be so drastic as to render their original character unrecognizable. Forging a new identity within Malaysia's changing socio-political landscape was therefore seen as important, even vital. It was nevertheless also seen as a nearly impossible task, because of differences between the desires of the previous Catholic hierarchy of the schools and the policies of the government: '... the government had reservations, whether we would be willing to commit our loyalty to Malaysia or whether we might be promoting policies that might not fit in ...' (Brother Peter).

With the exception of the former nun, Francesca, who argued that the role of the Church should not be to run 'elite schools for the rich', the government takeover of the mission schools post-1989 was against the wishes of all participants. This was reflected in a strong emotional backlash, most of it directed against the state: '... it's a loss to the country of an institution that produced some very good people ... and it could have, if allowed to continue, still generate people who were modern, progressive, egalitarian and fair to all....' (Robert). The disappearance of the missionaries was further seen as a generational gap and a loss to the current cohort of students, again reflecting very negatively on the government: 'The people felt united ... felt cared for ... I feel very angry, angry with the government for taking away the Missionaries' (Lai Mei).

The Question of New Identities

Most participants identified the schools' historically Catholic heritage as the last major anchor against significant change. As all participants had first-hand experience of the institutes prior to the dissolution of the Guild of Assisted Catholic Schools, they were asked if the schools could have valid 'post-Catholic' or 'non-Catholic' identities in light of changes since 1989. Their responses indicated that divorcing the schools from their religious origins would be difficult: 'Any religious school must have an identity tied to that religion. If not it becomes like any other school....' (Nathan).

Even respondents who felt a 'pluralism' connected to mission school identity didn't like the idea of suppressing the Catholic character of the schools due to non-Christian pressure:

> the [Catholic La Salle] Brothers said – 'majority of our students are [now] Muslims, so we'd like the chairman [of the Board of Governors] to be a Muslim' ... the Brothers are uniquely very pluralistic ... [however] ... Mission Schools ... have been targeted ... if [the others] had their way they would ... chop off crosses and many other things, but that's because of ignorance. (Faisal)

Indeed, even those who embraced an 'extension' of the schools' identity to include other religious or secular traditions preferred that a Catholic identity be the primary, guiding light. The rise in ethno-nationalism in Malaysia since the 1980s (Chew, 2000) was therefore identified as particularly detrimental and responsible for a 'loss of diversity' in the education system as a whole by the Muslim participants in particular.

Rather than being a footnote of history, participants described a palpable difference in the atmosphere and sense of belonging to the mission schools before and after their de facto nationalization in 1989. Only William and Dr Lawrence, who believed in the need for societal evolution, saw things differently: '... maybe until today, it would still be an identity with kind of a Western, colonial mindset. ... "Europeanism", more than anything ... Mission equals an identity with Westerners' (Dr Lawrence).

For some participants, equating mission school identity with Western colonialism placed a built-in expiry date on their inherited character, leading to a post-1989 heritage vacuum arising from the loss of control (but not legal ownership) by the Catholic church over the schools. This dealt a heavy blow to the schools' ability to continue claiming any special heritage: '... nobody knows what the identity is, because we're living in a period of flux... we are trying to forge a new identity ... but ... the "new" heritage has not been developed yet' (Dr Lawrence). Other participants, however, chose to frame their pro-Catholic stance in a more 'pluralistic' form: 'Heritage is whereby you can share your experiences with your children and your future generations ... [mission school heritage] ... it's very positive ... it's a good feeling ... the exchange of cultures, and then values' (Razak).

Unconverted 'Potential Energy': Heritage 'Continuity' and Tourism

The mission schools' old buildings were identified as their most important 'store' of unconverted 'potential energy' for tourism. Given that some schools have already been sold to developers and adaptively re-used as museums or retail outlets targeting tourists, some participants shared their opinions in kind: 'It would be so lovely if they were turned into museums, at least the structure would be there' (Emma). 'I would be happy, because knowing what the government has "progressed", at least the building, the façade would still be there' (Lai Mei). Such opinions highlighted an openness to tourism as an acceptable compromise between present-day usage patterns that do not adequately 'honour' the past and nostalgia for a vanished way of life. Participants suggested that the schools' old traditions were in fact more at risk if they continued to be run as government institutes where their residual 'missionary' character would eventually be replaced by something entirely different, and potentially Islamic, reflecting the majority Muslim student population of most of these schools in recent years.

Because the fading of mission school identity contributed to a strong sense of disillusionment, some participants identified responsible tourism via sensitive adaptive re-use of the heritage buildings as a way to preserve the authentic memories of their schools before a new 'other' – the current and later generations of alumni of the now nationalized schools – could make their alternative viewpoints felt. For many participants, any new narrative about the mission schools which ignores their foreign missionary links is itself ignorant and deserving of ridicule. The issue of 'mission school heritage', from their point of view, has grown from being a mere tradition into a fondly remembered ideology, now passing rapidly via telling and re-telling, into the status of a venerated *legend*.

The 'hollowing-out' of mission school identity was seen as a starting point for discussions concerning heritage, opening the door for alternative directions surrounding character, identity, and tourism in the future: 'I wouldn't mind [converting the schools for tourism], rather than they be used as government schools ... I would prefer that, you know, they remain as a legacy of the past' (Linda). Others suggested the inevitability of change due to the high cost of maintaining old buildings that no longer fulfilled their original purpose: '... there is a certain aura about them, the way they were built...[but] they just can't maintain these buildings anymore' ... 'if the school was run properly ... then I would give you all the support ... if you cannot do that, then I suppose it's better to turn it into a museum or some other place' (Francesca).

Some Catholic participants emphasized that converting the schools for tourism could in fact celebrate their role better than their evolution into standard government schools, by drawing a comparison with tourism in and around Jerusalem: 'Well you see, the Holy Land, the churches there are used for tourism ... I think it's okay' (Linda). One participant shared her experience of visiting a tourist-friendly Catholic monastery in Portugal as an example, comparing the situation there to the earlier presence of resident nuns in Malaysia's convent schools:

> I went to a monastery in Porto, there were some nuns there, they have to do a job ... to survive as well, they had to take care of the hotel, the restaurant ... I thought that it was good that the people were there still. It's a monastery. ... so, similar to the school. (Dr Aishah)

Discussion

The exchange of cultures and values highlighted as important by the research participants has strong parallels with how Pretes (2003) described heritage tourism attractions as 'hegemonic cultural producers'. It also equates with how Gonzales (2008) and Vecco (2010) argued that tourists are mainly interested in 'living an experience' and preferably a 'unique' one, highlighting the possibility of tourism being an acceptable vehicle to preserve mission school 'continuity'.

The interviews showed that the participants' desire to preserve mission school 'continuity' was of paramount importance, and their lack of success in this regard was a source of deep frustration. Rather than a replacement ideology, they identified a vacuum created by the removal of missionary staff and their humanist-Catholic ideology. This removal, largely blamed on the government, resulted in emotions ranging from anger and sadness to a grudging acceptance that change, however undesirable, is nevertheless inevitable. Without exception, the responses suggested an overriding desire that even if the present has been irrevocably changed, the past should be remembered in all its perceived glory. A CDA model based on the three-dimensional method developed by Fairclough captured the essence of the oral histories as follows (Figure 4.3):

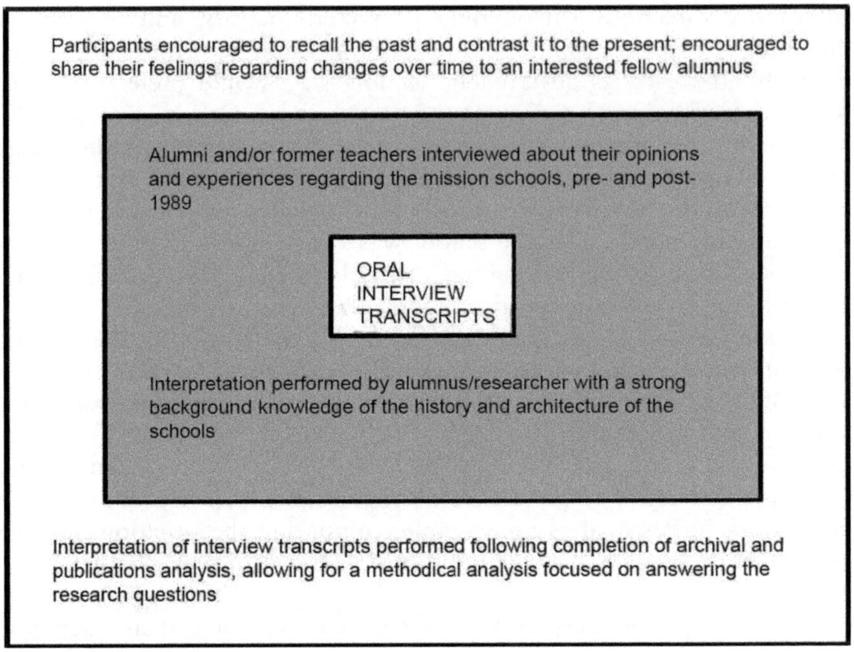

```
┌─────────────────────────────────────────────────────────────────────┐
│  Participants encouraged to recall the past and contrast it to the  │
│  present; encouraged to share their feelings regarding changes      │
│  over time to an interested fellow alumnus                          │
│   ┌───────────────────────────────────────────────────────────┐     │
│   │ Alumni and/or former teachers interviewed about their     │     │
│   │ opinions and experiences regarding the mission schools,   │     │
│   │ pre- and post-1989                                        │     │
│   │            ┌──────────────────┐                           │     │
│   │            │ ORAL             │                           │     │
│   │            │ INTERVIEW        │                           │     │
│   │            │ TRANSCRIPTS      │                           │     │
│   │            └──────────────────┘                           │     │
│   │ Interpretation performed by alumnus/researcher with a     │     │
│   │ strong background knowledge of the history and            │     │
│   │ architecture of the schools                               │     │
│   └───────────────────────────────────────────────────────────┘     │
│  Interpretation of interview transcripts performed following        │
│  completion of archival and publications analysis, allowing for a   │
│  methodical analysis focused on answering the research questions    │
└─────────────────────────────────────────────────────────────────────┘
```

Figure 4.3 A 'Three-dimensional CDA model' for narrative interviews

Figure 4.3 shows little ideological difference concealing relationships between the interview transcripts, discourse practice and sociocultural practice in the examination of narrative histories. This was partly because all interviews were conducted wholly in English, as requested by the participants themselves. Without any linguistic barrier, the narrative interviews therefore represent the 'closest' relationship of text, discourse and sociocultural practice found, in comparison to an archival review and publications analysis also conducted as a pilot for this study. Because most participants identified mission school heritage as belonging to history, many hoped for the possibility to celebrate and indeed even glorify it by re-creating a link between past and present. This again pointed to 'continuity' as the participants' strongest desire, because continuity, no matter how 'evolved', would require a respect for the schools' missionary traditions and, ideally, the participation in some way, shape or form, of the Catholic mission institutes themselves.

Preservation via Tourism-linked Change

The analysis of the interviews also showed that the 'erosion' of the mission schools' original ethos, spirit and character was not the result of their original philosophies being replaced by a new, nationalist narrative. Instead, it arose from a gradual 'asphyxiation' of the schools' character due to the removal of missionary headmasters, Catholic teachers, large numbers of Catholic students and, importantly, the phased removal of the English language as the dominant medium of instruction in Malaysia's public schools after 1970. Indeed, because Malaysia's once 'premier' mission schools have not managed to significantly influence the country's education system since 1989, they are now viewed mostly as part of its democratized (or, according to the more critical respondents, mediocre) system of public education.

Earlier academic studies suggest that this pattern could paradoxically enhance the appeal of the sites for tourism, because relics that have been abandoned or otherwise forgotten sometimes become more valuable than those in continuous use, as fragility renders them rare (Lowenthal, 1985). The current lack of touristic interest at Malaysia's functioning mission school sites shows, however, that long-term tourism cannot develop via nostalgia alone. Rather than just the 'inter-generational transfer of ideas' alluded to by Halbwachs (1980) and Southgate (2005), the findings show that for memory, identity and heritage to truly contribute towards an active touristic engagement, it is necessary for that 'inter-generational transfer of ideas' to become an 'inter-generational transfer of interest' to convert the 'touristic potential energy' of the mission school sites into active 'feet on the ground' tourism.

The emphasis on linking past to present in fact showed that the current 'lack of continuity' caused by government control of the mission schools was stoking an emotional conflict amongst the research participants, causing them to identify tourism as the most acceptable, even preferable bridge to re-establish continuity by honouring the missionary legacy in order to create the required 'inter-generational transfer of interest'. This inter-generational element is vital because the early mission schools are today mostly near or over a century old. They have passed the age when many relics risk being consigned to 'junk' status and discarded (Houston, 2013). Instead, the schools' heritage is judged all the more valuable because of its perceived fragility, creating a 'rarity premium' growing with the passage of time.

Drawing on his long teaching and later administrative experience in Malaysia beginning in the 1950s, Brother Peter confirmed that interest in protecting Malaysia's mission schools from a 'heritage', rather than 'educational' viewpoint only began in earnest after 1989. Before the dissolution of the Guild of Assisted Catholic Schools, they were appreciated mostly for their teaching, whereas from a purely cultural viewpoint they were perceived at best neutrally or sometimes even negatively due to their hierarchical, Euro-centric origins. Even from the Malaysian government's point of view, belated celebrations of mission school heritage in the 21st century are also now easier to accept because, as was argued by Lowenthal (1985), a past that is no longer fully 'alive' poses no real threat to the present, making the possibility of its re-interpreted revival easier.

Government involvement in the discursive re-construction of Malaysia's mission school heritage, whilst absent in the intangible sphere, is therefore evident in the granting of official 'heritage' status to many of the country's oldest schools. This 'tangible heritage' is often localized and community friendly, one of the few areas where the 'thematic interests' of the schools and the government coincide, equating the theme of 'heritage' again with the materialization of 'architecture'.

Architecture, Mission School Heritage and Tourism

Once an old building is defined as a heritage structure, there is an additional, natural tendency for the host city to want to show it off. In Malaysia, however, this often means sanitizing the past to remove elements deemed unattractive. In the case of the mission schools, this also often involves the removal of overtly religious symbolism from even legally protected buildings. As a result, wall-hung crucifixes, statues of Catholic saints and chapel altarpieces seldom survive the transition from church administration to government control. The same is often true for the school buildings that have been adaptively re-used for tourism. The building's original function and its historical authenticity can also sometimes struggle against the need to accommodate contemporary requirements for aesthetics, comfort and commercial viability, once again damaging the very relic a heritage designation is meant to protect.

At the same time, mission school architecture from before the Second World War mostly alludes to a romanticized memory of the colonial era. When new, their buildings allowed missionaries to maintain a sense of attachment to 'homelands' they had forsaken, leading to the contemporary perception of the schools having mostly 'colonial links'. This 'colonial baggage' is therefore a further, major consideration that cannot be ignored when adapting the mission school sites for tourism. Even in cases where that colonial heritage is to some extent 'suitable' to its context, as shown in Figure 4.4, constraints such as leasehold land titles can sometimes interfere with the ability of the mission institutes to exploit their old properties for tourism, trapping the buildings in a vicious cycle of government managed obscurity and missionary-induced poverty.

The dispersed locations of heritage mission schools in Penang, Ipoh, Taiping, the Cameron Highlands, Kuala Lumpur, Malacca and Johor Bahru in mostly unsympathetic local contexts has further complicated their promotion as viable tourist destinations, because they are often alone in increasingly incompatible modern surroundings. The exceptions to this rule have been at campuses that no longer function as schools or those in and around the UNESCO World Heritage sites of George Town and Malacca.

The combined hurdles of traffic congestion, student safety and economics, however, all serve to prevent functioning schools in general,

Figure 4.4 Main block of the Cameron Highlands Convent, built in 1935 on leasehold land, where the district council encourages the use of mock-colonial architecture (photograph by main author)

and mission schools in particular, from becoming conveniently accessible heritage attractions. Peak-hour traffic is in fact one of the main factors behind pressures to relocate Malaysia's inner-city schools. With few exceptions, most mission school sites are also not located within easy reach of train-based public transport, preventing easy access for casual tourists. Unlike the tourist-friendly old university towns of Europe, student safety is also an important barrier to unregulated tourist access at mission school sites, especially because many of the schools have primary as well as secondary students occupying the same extended campuses, making the average age of a mission school student well below the independent, 'young adult' category of a European university town.

Economics is a further barrier, because the government generally pays for the schools' running costs but not the physical maintenance of any buildings, despite the latter cost generally rising in tandem with a building's age. As confirmed by Brother Peter, this, and the wear and tear arising from visitor numbers, is something the mission institutes in Malaysia can ill-afford given their ageing and shrinking membership. The mission school sites are, therefore, even less viable for tourism compared to other 'public' buildings of a similar age, because they are 'public places' constrained by 'private ownership', yet without the typical

Figure 4.5 Former St Joseph's Novitiate in Penang (photograph by main author)

financial advantages of private property – such as the ability to charge entrance fees, gain rental income or monetize commercial branding.

All these advantages, however, can and do occur when the heritage of these sites is discursively re-constructed for tourism, which has so far occurred only following the closure of a school and the relocation of its students. When this happens, some former mission school sites have exploited their city-centre locations and flamboyant architecture to charge entrance fees, as in the case of neighbouring Singapore's original St Joseph's Institution campus, now the Singapore Art Museum. Others have exploited their distinctive architecture to serve as showpieces for new commercial developments, to 'stand out from the crowd' of otherwise similar competition, as in the case of the former Victoria Street Convent, also in Singapore, or the former St Joseph's Novitiate in Penang (Figure 4.5).

Now the centerpiece of the Gurney Paragon lifestyle mall, St Joseph's re-opened its doors in 2012, following its sale to a commercial developer and several years of adjacent construction, a full quarter century since the last missionary brother had been trained there. It serves as an example of how, in Malaysia, only when enough time has expired since the ending of the 'missionary presence', when most first-hand experience of teaching and learning has been forgotten, has the heritage of these places been fully exploited for tourism. The new owners even engaged the *Malaysia Book of Records* (November 2014 edition) to certify their new development as the '1st shopping Mall integrated with a Heritage Building' in the country, in order to obtain a 'superlative' branding advantage in Malaysia's increasingly competitive retail tourism market.

This points to the fact that although the custodians of mission school identity take pride in their intangible as well as their tangible heritage, recent evidence of re-branded properties suggests that the fading away of the 'living' component of that heritage has thus far been a prerequisite for meaningful, non-religious touristic activity to begin. The causes include psychological constraints, such as religious sensitivity and communalism, as well as physical constraints like access, safety and maintenance.

The overall effect of 'secular' tourism on the mission schools will probably depend on how many more years pass before their 'touristic potential energy' is converted into active tourism interest. If the transition occurs when the living memory of missionary teachers and alumni can still recall the 'glory days' of the schools' 'independent era', the effects of carefully managed heritage-based tourism could at best increase the pride of these stakeholders in their *almae matres* or at worst be seen as an unwelcome 'commercialization' of public spaces 'blessed' with a residual religious aura.

If the transition occurs many years from now, however, the emotive effects of tourism will probably be far lower and the overall effect welcomed as potentially more positive, because touristic activity might then be viewed as 'saving' these old buildings from any further 'defilement' arising from disinterested government control. Especially if the 'living memory' of the missionaries themselves eventually disappears from the population at large, a touristic transition might be particularly welcome at what would probably by then be the 'former' mission schools sites.

Conclusion

This research engaged a comparatively small group of people who consider themselves part of a larger group of 'custodians' of mission school heritage in Malaysia on account of their contribution to or experience of teaching and learning in these schools during their time as autonomous institutes managed by the Catholic church before 1989. The research could not identify a meaningful counter-narrative regarding mission school heritage from Malaysia's federal government despite its de facto control of these schools in recent decades. It is therefore the voice of a subordinate group controlling a heritage narrative by connecting a celebrated past to an obscure present.

By highlighting the touristic potential of Malaysia's mission schools, this study adds to the relatively limited body of knowledge regarding heritage sites that are not yet functioning tourist attractions. It suggests that a discursive 'heritage re-construction' is sometimes necessary to 'convert' a site's 'touristic potential energy' into active tourism interest. It provides suggestions for how plural societies can utilize tourism as a vehicle for greater inter-communal appreciation of 'alternative history' even in environments when these 'alternative narratives' are not officially recognized, sanctioned or celebrated. In other words, the production and promotion of colonial heritage for tourist consumption, even if controversial, may act as a catalyst for nation building and social cohesion, especially in plural societies like Malaysia. In this respect, celebrating a common colonial past and promoting it to tourists could also be conceived as a social strategy to minimize internal conflicts among the different Malaysian ethnicities.

Ultimately, for Malaysia's mission schools, a conversion of function away from education towards tourism is in fact perceived by many as an opportunity to preserve the heritage and 'authenticity' of these sites to a greater degree than would occur by maintaining their original function 'under different management'. This 'logical paradox' of 'preservation via change' adds to the heritage tourism industry's discussions about 'conservation', 'adaptive re-use' or 'modern replicas' of historical buildings by identifying 'change' as a useful tool to convert touristic potential to active tourism interest.

References

Amerian, M. and Esmaili, F. (2015) A brief overview of critical discourse analysis in relation to gender studies in English language textbooks. *Journal of Language Teaching and Research* 6 (5), 1033–1043.
Ashworth, G.J. and Tunbridge, J.E. (1996) *Dissonant Heritage*. Chichester: Wiley.
Butler, G., Khoo-Lattimore, C. and Mura, P. (2012) Heritage tourism in Malaysia: Fostering a collective national identity in an ethnically diverse country. *Asia Pacific Journal of Tourism Research* DOI:10.1080/10941665.2012.735682.
Chew, M. (2000) *The Journey of the Catholic Church in Malaysia 1511–1996*. Kuala Lumpur: Catholic Research Centre.
Chouliaraki, L. and Fairclough, N. (1999) *Rethinking Critical Discourse Analysis*. Edinburgh: Edinburgh University Press.
Coon, H., Carey, G., Fulker, D. and DeFries, J. (1993) Influences of school environment on the academic achievement scores of adopted and nonadopted children. *Intelligence* 17, 79–104.
Desforges, L. (2000) Travelling the world: Identity and travel biography. *Annals of Tourism Research* 27 (4), 926–945.
Fairclough, N. (2001a) *Language and Power* (2nd edn). Harlow: Pearson Education.
Fairclough, N. (2001b) Critical discourse analysis as a method in social scientific research. In R. Wodak and M. Meyer (eds) *Methods of Critical Discourse Analysis* (pp. 121–138). London: Sage.
Gonzalez, M. (2008) Intangible heritage tourism and identity. *Tourism Management* (29), 807–810.
Halbwachs, M. (1980) *The Collective Memory*. New York, NY: Harper & Row.
Hall, C.M. and Tucker, H. (eds) (2004) *Tourism and Postcolonialism: Contested Discourses, Identities and Representations*. Abingdon: Routledge.
Hall, C.M. and Lew, A.A. (2009) *Understanding and Managing Tourism Impacts*. Abingdon: Routledge.
Henderson, J.C. (2004) Tourism and British colonial heritage in Malaysia and Singapore. In C.M. Hall and H. Tucker (eds) (2004) *Tourism and Postcolonialism: Contested Discourses, Identities and Representations* (pp. 113–125). Abingdon: Routledge.
Hollinshead, K. (2004) Tourism and new sense: Worldmaking and the enunciative value of tourism. In C.M. Hall and H. Tucker (eds) (2004) *Tourism and Postcolonialism: Contested Discourses, Identities and Representations* (pp. 25–42). Abingdon: Routledge.
Houston, D. (2013) Junk into urban heritage: The neon boneyard, Las Vegas. *Cultural Geographies* 20 (1), 103–111.
International Council on Monuments and Sites (1994) *The Nara Document on Authenticity*. Nara: ICOMOS.
Kuipers, M. and Ashworth, G. (2001) Conservation and identity: A new vision of pasts and futures in the Netherlands. *European Spatial Research Policy* 8 (2), 55–65.
Lim, K.S. (2007) 48 hrs for two BN MPs to apologise for slur on mission schools. Extracted from: https://blog.limkitsiang.com/2007/12/05/48-hrs-for-two-bn-mps-to-apologise-for-slur-on-mission-schools/.

Lowenthal, D. (1985) *The Past is a Foreign Country*. Cambridge: Cambridge University Press.
Meyer, M. (2001) Between theory, method and politics: Positioning of the approaches to CDA. In R. Wodak and M. Meyer (eds) *Methods of Critical Discourse Analysis* (pp. 14–31). London: Sage.
Mura, P. and Wijesinghe, S.N. (2019) Behind the research beliefs and practices of Asian tourism scholars in Malaysia, Vietnam and Thailand. *Tourism Management Perspectives* 31, 1–13.
Palmer, C. (1999) Tourism and the symbols of identity. *Tourism Management* 20 (3), 313–321.
Pappu, C. (1996) *Malaysian Catholic Schools at the Crossroads*. Manila: De La Salle University.
Perks, R. and Thomson, A. (2006) *The Oral History Reader* (2nd edn). Abingdon: Routledge.
Pretes, M. (2003) Tourism and nationalism. *Annals of Tourism Research* 30 (1), 125–142.
Southgate, B. (2005) *What Is History For?* Abingdon: Routledge.
Tan, K.K.H. (2011) *Mission Schools of Malaya*. Subang Jaya, Malaysia: Taylor's Press.
Tan, K.K.H. (2015) *Mission Pioneers of Malaya*. Subang Jaya, Malaysia: SABD.
Tan, K.K.H. (2017) *Heritage Re-construction and Tourism: The Multiple Narratives of Malaysia's Mission Schools*. Subang Jaya, Malaysia: Taylor's University.
Throsby, D. (2001) *Economics and Culture*. Cambridge: Cambridge University Press.
Vecco, M. (2010) A definition of cultural heritage: From the tangible to the intangible. *Journal of Cultural Heritage* 11 (3), 321–324.
Ward, I. and Miraflor, N. (2009) *The Tradition, the Legacy, the Future*. Ipoh, Perak, Malaysia: Media Masters Publishing.
Wijesinghe, S.N., Mura, P. and Culala, H.J. (2019) Eurocentrism, capitalism and tourism knowledge. *Tourism Management* 70, 178–187.

5 Empowering Package Tour Travellers by Disempowering Tourism Operators? Assessing the Effectiveness of the Tourism Law of China

Nan Chen, Kevin Burns and Jing Wang

Introduction

The Tourism Law of the People's Republic of China (the Chinese Tourism Law 2013, abbreviated as 'the CTL2013' in this chapter) directed at empowering and protecting package tour travellers (PTTs) came into force commencing on 1 October 2013. This was the first national tourism law to be enacted in China, focusing on low-fare package tours, the unsustainable development of tourism resources, and uncivilized tourist behaviour. This chapter will examine the effectiveness of this law in protecting tourist benefits by governing low-fare package tours, which remain in a chaotic state.

The term 'low-fare package tours' (LFPTs) originates and is almost exclusively used in the Chinese context (Chen et al., 2011), describing a type of all-inclusive package tour incorporating all the aspects of a trip (i.e. transportation, accommodation, meals, sightseeing, and tour guides) yet charging an unreasonably cheap price. LFPTs have been extremely popular for more than two decades among Chinese travellers who are price sensitive. LFPTs always come with a catch: there is no such thing as a free lunch. Tourists are taken to many designated shops and requested, sometimes forced, to buy overpriced products (Ap & Wong, 2001). Tour operators, especially the tour guides (TGs) and drivers, can get a substantial handshake from the shopping to offset the loss on the tour cost. To urge LFPT participants to purchase as much as possible, various unethical means, even violations, have been applied by TGs, such as coaxing, misleading or deceiving, coercing, even aggression. Negative news about LFPTs are emerging constantly, such as LFPT participants who refuse to make a purchase often being abused emotionally or physically by TGs; the entire group is abandoned in the middle of the trip because the participants do not comply with the hidden rules of LFPTs (Wang et al., 2015). The rise of social media has accelerated the circulation of such negative reports on LFPTs, seriously jeopardizing the reputation of the entire tourism industry and sparking outrage among consumers towards tour operators, even tourism administrative departments (Chen et al., 2018). To regulate the operation of travel agencies and to protect the rights and interests of tourists, the China National Tourism Administration (CNTA) began drafting the first tourism law in 2009 under the call of various stakeholders (Tang, 2017). The CTL2013, therefore, explicitly forbids tour operators from organizing tourism activities at unreasonably low prices, from specifying shops, and from imposing additional fee-charging programs to 'powerless consumers' (see Articles 9 and 35). Briefly, the CTL2013 aims to empower PTTs by disempowering tourism operators.

Nevertheless, reports from the field argue that the encounters and experiences of PTTs have deteriorated even further since the implementation of the CTL2013, indicating that the law has not been enforced at the operational level (e.g. Fan & Xu, 2016; Luo et al., 2017). At the same time, increasingly tour operators have begun questioning whether the CTL2013 has resulted in 'over-regulation' – some of them believe LFPTs are the natural consequence of free-market competition and thus should not be regulated by administrative power. Such a debate on 'over-regulation or under-regulation' is not new and has long continued in the market economy. Sunstein (1990) revisited this regulatory paradox by advising that government regulation is indispensable to counteract breakdowns in the private market, even though the regulatory programs are not always successful. Therefore, attention to the actual consequences and impact of regulations can provide valuable insights for both legislators and administrators.

The effectiveness of the CTL2013 in empowering PTTs is also controversial among tourism scholars: both the protected/empowered party (i.e. PTTs) and the regulated/disempowered party (i.e. tour operators) have raised various doubts and complaints about the implementation of this law (Yang & Tan, 2016; Yin & Zhu, 2017). Tour operators frequently complain that this law is biased towards PTTs (Wang et al., 2018); TGs have claimed that this law does not protect them at all (Xu & McGehee, 2017); while PTTs have reported that this law is an empty shell for them (Fu & Jue, 2015). This chapter is based on exploratory research which aimed to examine the effectiveness of the CTL2013 from the perspective of both the empowered and disempowered parties. The chapter thereby identifies the underlying paradoxes pertaining to the CTL2013, particularly in relation to its perceived effectiveness and ineffectiveness.

Low-fare Package Tours: Treat or Trap?

The package tour business has dominated the international travel market for decades and remains the largest segment of the travel industry worldwide. Group package tours (GPTs) are a type of organized tour where members travel in a group, pay for a bundle of travel services, including airfares, accommodation, meals and transport, and are escorted by a guide (Dwyer *et al.*, 2007). It is GPTs that make tourism consumption both accessible and affordable to a huge number of mass tourists and facilitate medium and then long-haul international travel (Cohen, 1972). The dominance of GPTs can be attributed to their economic advantages (i.e. value for money and reasonable prices) over independent travel. For some tourists, purchasing a package tour is more valuable, time efficient, and convenient than buying a range of separate tourism products and services and it reduces uncertainties and perceived risks (Kim *et al.*, 2009). The most preferred mode of travel for Asian tourists, particularly Chinese tourists, is still GPTs – especially when travelling abroad, because of its convenience, economic pricing, and reduced language barriers (Chen *et al.*, 2018).

However, the experiences of contemporary Chinese GPT participants tend to be influenced and even constrained by the practice of commission tours, where tour operators offer very low-fare tours and try to reap profit through commissions from shopping and from extra-paid entertainment activities (Zhang *et al.*, 2009). This unique type of GPT is often called a 'zero-dollar tour' or 'zero/minus-fare tour' and is touted as a great holiday at a fire-sale price (Dwyer *et al.*, 2007; Jia, 2006). Chen *et al.* (2011) proposed an analytic framework portraying the three fundamental elements of 'zero-fare' GPTs in the outbound tourism market: namely, low price in the source market, inferior quality at the destination, and the 'zero-fare' relationship between the outbound tour operators (OTOs) and the inbound tour operators (ITOs) who jointly operate LFPTs. The 'zero-fare' relation highlighted as the core of LFPTs leads to lower prices in the source market and inferior quality at the destination.

The operating mechanism of LFPTs is illustrated in Figure 5.1, comparing them with the business model for GPTs with a reasonable price. 'Zero-fare' GPTs indicate that OTOs transfer no tour fares downward to ITOs, thus ITOs rely heavily on commissions and kickbacks from tourists' extra expenditures at the destination. The 'zero-fare' relation is remarkable in contrast to the regular business model of GPTs, in which OTOs and ITOs share tour fares on a pre-negotiated proportional basis (Chen *et al.*, 2011; Jia, 2006). Despite

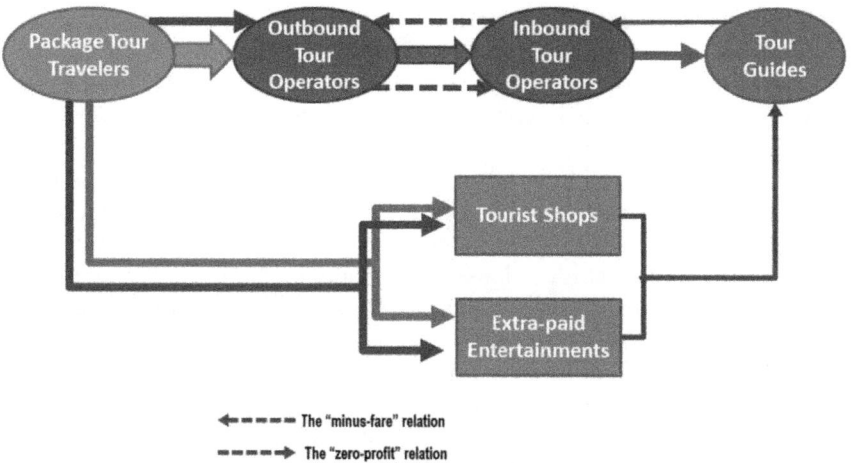

Figure 5.1 Business models for normal GPTs and LFPTs (adapted from Jia, 2006)
Note: The direction and weight of the arrows represent the flow and the magnitude of tour fares. The outer line illustrates the business model of LFPTs; the two solid arrows between Outbound Tour Operators and Tour Guides do not exist in the LFPT mode.

being an abnormal product of the vicious competition among tour operators (i.e. a price war), LFPTs have shown a robust vitality by surviving several tourism consumption recessions. For instance, LFPTs in Thailand have persisted despite a dozen measures initiated by both Chinese and Thai authorities (Zhang *et al.*, 2009).

Traditional GPT business has exposed a variety of quality problems, such as misrepresentation of information, tour operators defaulting on the contract, falsification of product information, and deceived consumption, i.e. deliberately misleading the tourist (Kim *et al.*, 2009). The inferior quality problems of LFPTs are even more prominent – not just cheating tourists but even involving aggressive or illegal actions, such as threatening and locking PTTs inside the designated shops to force sales (Wang *et al.*, 2015), which infringe personal freedom. Having seen first hand the damage caused by LFPTs to participating tourists, destination image and industry reputation, many tourism scholars have examined the causes and mechanism of this unique GPT to seek a corresponding antidote. For example, Jia (2006) identified a wide range of external and internal reasons underlying the prevalence of LFPTs, including Chinese consumers' limited rational decision-making, information asymmetry between PTTs and tour operators, GPT homogeneity and intensive competition among tour operators, and low business ethics. Prideaux *et al.* (2006) clarified the origin and development of LFPTs in Asian markets and indicated an enigma for ITOs – to increase GPT prices or to eliminate business practices? The authors believed government intervention (e.g. licensing, and penalizing TGs and ITOs who participate in LFPTs) was necessary, given the irreconcilable conflicting interests between service producers/providers and consumers, as well as between OTOs and ITOs. Moreover, Zhang et al. (2009), in seeking the causes of zero-fare GPTs, identified nine factors, including lopsided terms in contracts, the dominating bargaining power of tour agencies (TAs), and government deregulation of the industry. LFPTs are recognized as a carefully fabricated scheme involving a wide range of interested parties, that breaches consumer rights and business ethics, and which thus should be regulated effectively even if it cannot be eradicated.

Although many studies have been conducted on LFPTs, most of them have focused on their origin, development, and impact, and they have often failed to identify their causes and the factual features (Chen et al., 2011). Moreover, little attention has been paid to the legitimacy and regulatory issues of LFPTs (Tse & Tse, 2015). Importantly, extant research, particularly in the management field, has employed stakeholder theory to identify the interests of all parties within and outside the industry. However, as stakeholder theory seminar author Goodpaster (1991) pointed out, the paradox of stakeholder theory is that it treats stakeholders as both the means to the ends and the ends in themselves. Stakeholder theory postulates that the voices of all the stakeholders should be heard while making a decision, regardless of the power or interest held by stakeholder groups (Laplume et al., 2008). But Robson and Robson (1996) advocated that the real interest balance and a moral agreement among stakeholders can never be attained owing to inequalities in power and knowledge. In the current context, the enactment of the CTL2013 has actually laid down the 'correct' ways of developing tourism from the perspective of the government (Tse & Tse, 2015), thereby compelling the stakeholders involved in LFPTs to reach an agreement that tourist benefits are under priority protection. The perception of this priority is thus one of the main predictors for stakeholder support for the law, which is a necessary precursor to law enforcement. The paradox in this study is that managers should consider the interests of all stakeholders and not just shareholders, those earning revenue from LFPTs. Previous research on the tourism law was mostly interested in the protected party – the tourists; however, the exclusion of regulated parties may pose obstacles toward implementing the CTL2013. This study, therefore, examined the effectiveness of the CTL2013 from the perspective of both the empowered (i.e. PTTs) and disempowered parties (TAs and TGs).

Methodology

In order to explore the opinions of the three paramount parties involved in the LFPTs (i.e. TAs, TGs and PTTs) with regard to the effectiveness of the CTL2013 in empowering PTTs by disempowering tour operators, an exploratory interpretive approach was adopted. Through in-depth interviews conducted within the constructivism-interpretive research paradigm, our aim was to gather socially constructed knowledge in order to better understand the meanings that people construct in the particular social context under study (Jennings, 2005). The interview protocol was semi-structured, with a flexible agenda to focus the interviewees' responses and to encourage open and free-flowing dialogue (Jennings, 2005). The interview questions for TA managers and TGs focused on their working experience with and perceptions of LFPTs, as well as the perceived influences of the CTL2013 on their jobs and the business of TAs. The questions for PTTs focused on their participation experience with, or perceptions of, LFPTs, and their knowledge of the CTL2013 and its perceived effectiveness in empowering them to protect their rights and interests.

A tour guide training expert was employed as the interviewer because he could easily access to both front-line staff (i.e. TGs) and TA managers based on his professional network. Moreover, as the research topic was deemed sensitive, it was considered important that the interviewer would be regarded as an 'insider' by interviewees, thus enabling the building of mutual trust. The recruitment of interviewees came to an end when information saturation was reached. A total of 29 qualified respondents were interviewed in the first quarter of 2016. Interview transcripts were analysed following an inductive logic since the potential subject themes were expected to be drawn from the raw interview data, rather than from prior knowledge or theories. Therefore, the coders followed a standard procedure of 'open coding – creating categories – abstraction' to analyse the interview data (Elo & Kyngäs, 2008). To ensure the accuracy and confirmability of the coding results, two coders (native Chinese speakers who spoke fluent English) with different backgrounds analysed the interview transcripts independently and developed two coding sheets. Differences between the two sheets were noted and discussed amongst the coders and the interviewer. After several iterations, a consensus was achieved. The trustworthiness of this research was established through investigator and general triangulations (Decrop, 1999). The thick description of the findings and discussion in the following section also contributed to the credibility and dependability of this study.

Interestingly, 85% of the interviewed PTTs and one-third of front-line tour guides/leaders demonstrated limited or superficial knowledge of the CTL2013. More than half of the PTTs only learned from mass media (including internet and TV programs/news) and acquaintances that the CTL2013 was launched to ban LFPTs and forced shopping. Therefore, they had to rely more on mass media and social communications than the other two groups of respondents, as well as the personal travel experience with LFPTs if they had it, to obtain relevant information about the implementation of the law and to make their own judgments. The industry practitioners (i.e. TA managers and TGs) evaluated the implementation effectiveness of the CTL2013 mainly according to their own working experience and observations. Demographic characteristics (see Table 5.1) did not seem to have relevance to their attitudes. The

Table 5.1 Interviewee profiles

No. of tour operators	Name	Gender	Age	Education level	Job position	Working experience	Annual income (RMB 10,000)
1	CSP	Female	22	Undergraduate	Part-time tour guide	8 months	7
2	FYC	Male	26	Undergraduate	Tour guide	5 years	6
3	ZYL	Female	30	Undergraduate	Tour leader/escort	6 years	8
4	SYS	Male	29	Undergraduate	Tour guide	6 years	16
5	CRC	Male	33	Undergraduate	Tour guide	10 years	12
6	TY	Female	36	Undergraduate	Tour guide	12 years	7
7	WZS	Male	35	Undergraduate	General manager	12 years	9
8	CXJ	Male	43	Undergraduate	General manager	20 years	20
9	ZBQ	Male	52	Undergraduate	General manager	25 years	28

No. of PTTs	Name	Gender	Age	Education level	Occupation	Previous experience with low-fare GPTs	
1	ZX	Male	22	Undergraduate	College student	No	
2	LX	Female	21	Undergraduate	College student	Yes	
3	ZY	Female	53	High School	State-owned enterprise employee	No	
4	ZC	Male	36	Undergraduate	Civil servant	Yes	
5	LY	Female	45	Undergraduate	Joint venture staff	No	
6	WX	Female	21	Undergraduate	College student	Yes, unintentionally	
7	LZ	Male	52	High School	Dormitory administrator	Yes, like it	

(*Continued on next page*)

Table 5.1 (Continued)

No. of PTTs	Name	Gender	Age	Education level	Occupation	Previous experience with low-fare GPTs
8	XY	Female	45	Undergraduate	Self-employed	Yes, intentionally
9	ZD	Male	65	High School	Retired	No
10	DX	Female	41	Undergraduate	Teacher	No
11	YX	Male	47	Undergraduate	Joint venture staff	Once, dislike it
12	JX	Female	32	Postgraduate	Foreign enterprise employee	No, unreliable
13	LF	Female	23	Undergraduate	College student	Once
14	MJ	Male	56	Undergraduate	Businessman	No
15	ZN	Male	53	Postgraduate	Teacher	Yes, very dissatisfied
16	WJN	Female	25	Undergraduate	Private enterprise employee	Yes, dislike it
17	SXF	Female	40	Undergraduate	Joint venture Staff	Yes, experienced a conflict
18	QH	Male	57	Undergraduate	Civil servant	Once, terrible
19	LWQ	Male	41	Undergraduate	Private enterprise employee	Once, terrible
20	FY	Female	58	High School	Self-employed	No

following section presents the findings in relation to the perceived effectiveness and ineffectiveness of the CTL2013.

Perceived (in)Effectiveness of the CTL2013

The respondents' evaluation of the effectiveness of the CTL2013 can be categorized into three types: effective, limited/partial effective, and ineffective; among which the perception of ineffectiveness (n = 12) overwhelmingly outnumbered perceived effectiveness (n = 7) or limited effectiveness (n = 10). Their evaluation of the effectiveness of the CTL2013 is normally determined by their perceptions of LFPTs. If the respondents viewed LFPTs as reasonable, reflecting market demand, they normally embraced the CTL2013 in regulating and restricting TAs' operations and TGs' performance. However, if the respondents regarded LFPTs as an abnormality of the tourism market, which should be eliminated immediately by administrative power, they generally felt very disappointed with the effectiveness of the CTL2013.

The preliminary effectiveness of the CTL2013 is demonstrated by the reduction in LFPTs and the increased number of reasonably priced and high-end GPTs (Ma *et al.*, 2015; Tse & Tse, 2015). Many small TAs were closed down because they could not survive without LFPTs. For consumers, the reduction in LFPTs meant that the average price of GPTs with the same itinerary was raised significantly. The price rise may inhibit some potential consumers' purchase desire, but no-shopping GPTs with higher service quality are able to attract more upscale consumers who possess a higher level of education and wealth. However, a consequence of this is suspected price discrimination: consumers who cannot afford the normal GPT price may be completely deprived of travel opportunities. From their perspective, the CTL2013 did not empower but actually disempowered some potential PTTs.

Additionally, new changes in product description and to the travel service contract were recognized by both PTTs and TGs. The CTL2013 requests TAs to provide more specific details in the description of GPT products, especially in the itinerary, to protect PTTs' right to know. As one TG said:

> In the past, if TGs want to change the itinerary and add designated shops during the trip, we only need to achieve an oral agreement with the GPT participants; but now all tourists must sign a written agreement ... The standard contract requires TAs must sign a series of supplementary agreements with GPT participants which indicate specific arrangement and service criteria, or potential changes, such as the number and order of sightseeing sites, the number of dishes in group meals, and the number of seats on tour buses. (Mr SYS)

Moreover, the implementation of the CTL2013 has successfully increased the legal awareness of general consumers, has regulated illegal operations by TAs, and protected TGs' rights to some degree. More than half of the tourists expressed their trust in the CTL2013 – they believed this law was indeed trying to empower and protect PTTs, a relatively disadvantaged group, rather than tour operators – and thus will be very useful in litigation related to tourism consumption. For example, one tourist considered the CTL2013 as 'a deterrent to TAs and TGs who are trying to infringe rights and interests of PTTs'; and another one believed the law can 'resist unreasonable or illegal phenomena by increasing media exposure and raising public awareness'.

Similarly, although all TGs unanimously believed that the CTL2013 is mainly concerned with tourist rights and interests, they all embraced the implementation of this law because its regulation on LFPTs has also indirectly protected TGs' rights and interests. For example, law enforcement has exempted them from becoming involved LFPTs, at least temporarily. None of the interviewed TGs liked LFPTs, because most TGs engaged in unethical behaviour (e.g. forcing LFPT participants to purchase or tip) under huge economic and psychological pressure, even a security risk from conflicts with tourists. Moreover, Article 38 requests TAs to sign labour contracts with the TGs they employ, pay remuneration and social insurance premiums for them; and for part-time TGs, service fees need to be paid to the full amount. It can be said that the CTL2013 has guaranteed the income of TGs by restraining the operation of TAs. To further improve TGs' income, CNTA even suggested encouraging voluntary tipping for high-quality service given by TGs. The TA managers also recognized the effectiveness of the CTL2013 in improving TGs' allowances and enhancing their basic salary, as well as in leading to more no-shopping GPTs with higher service quality. However, TGs also stated that the commission they could obtain from shopping and extra-paid entertainments had been reduced significantly, due to the reduction in LFPTs and the decline in the number of PTTs. Although they are more likely to receive a higher service charge and tips from upscale PTTs, unknown long-term gains cannot compensate for the losses that have occurred. Therefore, TGs normally showed a contradictory emotion towards the effect of the CTL2013: they were not happy with the decline in income caused by the implementation of the law, but they were glad to see a more regulated tourism market with fewer LFPTs. This creates another paradox for the implementation of the law: should we sacrifice short-term benefits to pursue a long-term benefit or vice versa?

In addition, more than half of the respondents believed that the launch of the CTL2013 is better than nothing. They have recognized its function in regulating the disordered GPT market and TAs' operation – for instance, the improved travel service contracts and fewer consumption traps. As one TA manager (Mr FYC) stated: 'Compared to the chaos in previous years, the situation has gradually improved ... both tourists and TAs are looking for a new balance of interests.'

Briefly, the law has successfully curbed the rampant growth of LFPTs, although it may only be temporary. The partial or temporary effects of the CTL2013 were emphasized by many respondents, for the following reasons. First, the power of this law is limited – it cannot regulate overseas TAs (i.e. ITOs) that offer TGs commission through shopping and extra-paid entertainments. Second, the law lacks specific advice and practical measures for enforcing agencies, thus leading to loose supervision and to poor implementation. Many consumption traps in LFPTs still exist – they have just become more concealed than before. Third, combating LFPTs takes time – maybe three, five, or 10 years. It is unrealistic to think one can eliminate LFPTs immediately. Fourth, implementation of the law requires close collaboration among multiple departments but joint enforcement in reality is very challenging. For all of the above reasons, the effect of the CTL2013 was regarded as unsustainable.

Contrarily, about 40% of the respondents believed that the CTL2013 was ineffective in its actual implementation. The evidence they offered includes endless negative news reports of disputes between TAs/TGs and LFPT participants; the fact that LFPTs were still available in the market; and consumption traps did not diminish but became more concealed and deceptive. As one TG stated:

> At the beginning of implementation, the law is effective in deterring small TAs' unethical operation. Most TAs stopped the provision of LFPTs temporarily because they were not sure about the benchmark of 'low-fare'. Consumers were confused by the abrupt unavailability of familiar LFPTs and the rising price of GPTs with the same itinerary. For example, some tourists visit HK and Macau only for shopping and they have been used to joining a LFPT. When the law just came into effect, we had to tell potential customers that LFPTs were not available anymore, thus lost these customers. Then, the standard contract issued by the government allows shopping again; the market returned to chaos. Consumers became even more confused and disappointed – they just learned about the essence of LFPTs through the publicity and education concerning the CTL2013, but the illegal LFPTs reappeared. (Ms TY)

Similar to the above statement, almost all TA managers and TGs identified the main culprit that encouraged TAs to stop watching and revive LFPTs: the standard contract of travel service jointly issued by the CNTA and the State Administration for Industry Commerce. This contract unintentionally inspired TAs to get around the law and restore shopping and extra-paid activities through the signing of supplementary agreements with PTTs. As a TA manager (Mr ZBQ) revealed, some TAs, especially those that were small in size, evaded the law by taking advantage of the contract addendum to carry out illegal operations and keep LFPTs:

> Now, PTTs are requested to sign two supplementary agreements – the 'Agreement on extra-paid activities' and the 'Agreement on Shopping'

when signing the contract with TAs. Because the CTL2013 stipulates that shopping and optional entertainment activities cannot be included in the itinerary, TAs do not dare to show these items on a formal contract. To meet the demand of consumers for shopping and attractive entertainments and retain the profit, TAs urged the government to issue a model contract that allows them to include shopping and optional activities again in supplementary agreements. Since then, PTTs have been requested to sign several separate agreements with TAs besides the contract; if you don't sign, you cannot join the pre-designed GPTs containing shops and extra-paid activities. PTTs actually lost their rights to choose ... Article 35 shows a severe legislative intervention in the free market; this is excessive administration and ultra-vires. (Mr ZBQ)

It can be seen that TAs have boldly come up with various countermeasures to evade the law, under the encouragement of the standard contract. The widely adopted contract addendum not only failed to protect tourist rights and interests but also led to the revival of LFPTs. Just like putting old wine in a new bottle, signing additional agreements with every PTT has wasted even more social resources. All interviewed TA managers' evaluation of the CTL2013 was quite low, even though they indicated a supportive and optimistic attitude towards the launch of the law. They agreed that contemporary tourism development in China does require national law to empower and protect tourists, to regulate unethical operations of the travel-related industries; but the CTL2013 has been shown to be ineffective in prohibiting LFPTs and forced shopping in the practice, or at least much less effective than expected. Similarly, many tourists reported that no significant improvement in the service quality of TAs had been observed according to their own, or their acquaintances', travel experiences with GPTs. For example, tourists believed that the CTL2013 was ineffective in resisting opaque consumption. They did not identify any essential changes in the tourism market – LFPTs are still available; TAs are still promoting LFPTs just not that blatantly; and TGs are still taking PTTs to designated shops by subtle means, such as deliberately dropping tourists off at a specific shopping area in the name of 'free activity'.

To conclude, the interviewees in our study believed that the effectiveness of the CTL2013 was diminishing over time: it was effective in the first half to one year of coming into effect, because most tour operators were maintaining a wait-and-see attitude towards the implementation of the law and its actual impact on the tourism market. They wished to decide whether to retain LFPTs or not through careful observation over some time after the law's enforcement – the final decision will depend on the intensity of supervision of the relevant administrative departments (e.g. TAOs) over the TAs' business, as well as the penalties imposed for specific cases of violation. During this wait-and-see period, most TAs have restrained themselves from operating

LFPTs openly and unreservedly. However, after a period of observation, TAs found the enforcement of the law was unfavourable, and thus started operating LFPTs again but by different approaches. Especially after the issue of the standard contract, TAs craftily added contract addendums to delimit the scope of their rights and obligations. PTTs were thus disempowered again – they had to continually bear with the GPTs with designated shops and extra-paid entertainments if they wanted to join a GPT. The tourism market returned to a state of chaos again. Therefore, we can say that the effectiveness of the CTL2013 was diminished soon after the issue of the standard contract. In a word, the effectiveness of the CTL2013 has been short lived and very limited from the perspective of all three types of stakeholder.

Paradoxes Influencing the Effectiveness of the CTL2013

Even though the stringent but principle-based regulation yielded no protection of PTTs in many settings, the CTL2013 is not a self-defeating regulation because all the stakeholders we interviewed agreed that the enactment of the CTL2013 is better than nothing. Four major entities are believed to account for the unfavourable effects of the CTL2013: (1) Governments and administrative departments, including the lawmakers, the CNTA and the local TAOs at both provincial and municipal levels, received the most criticism, followed by (2) PTTs. Then (3) TAs and (4) TGs also have an unshirkable responsibility for the notorious LFPTs and the poor implementation of the CTL2013. Thus, CTL2013 reflects ineffective political prowess, a stereotype threat faced by tour operators and ultimately the self-disempowerment of consumers.

Paradoxes in the principles-based regulation approach

The ambiguity in the key terms and articles of the CTL2013 may imply a 'principle-based regulation' approach adopted by the legislators, as opposed to a 'rules-based regulation', which relies upon detailed, prescriptive rules. Principle-based regulation relies more on high-level, broadly stated rules or principles to set the standards by which regulated corporations must conduct business (Black, 2008). However, this regulatory approach is not that effective in regulating the contemporary chaotic GPT market of China, because of the following paradoxes – some of which are inherent (Black, 2008) while some arise from enforcement of the CTL2013.

(a) Interpretive and communicative paradoxes

The imprecise terms, expressed in ordinary language (e.g. LFPTs, unreasonably low price), may facilitate understanding and communication of the regulatory objectives as well as the responsibilities of tour operators more clearly for ordinary consumers, but they may also impede other people's interpretation and communication, particularly the regulated practitioners and law enforcers (Black, 2008). Ambiguous or inaccurate legal terms may receive various specific interpretations from different stakeholders because every stakeholder intends to interpret for their own sake. Legislators were widely criticized for being ivory-towered and for detaching themselves from industry practice, thus ignoring the realistic market demand and industry operations. An insufficient number of tourism professionals, including academia and industry practitioners, were involved in the legislation process. A senior TA manager (Mr ZBQ) we interviewed who had participated in the consultation on the law, revealed that the legislators of the CTL2013 did not conduct an in-depth and comprehensive investigation into the GPT market nationwide, nor did they involve all stakeholders in the legislation. Consequently, their limited or inadequate understanding of the GPT market caused ambiguity in the definition of relevant terms and left various loopholes in the CTL2013, which also led to poor operability. For example:

> The lawmakers showed their ignorance of the industry practice. They just sat back and pontificated; they made the law behind a closed door. How can you simply adopt 'zero/negative/unreasonable low-fare', which are the jargon used by industry practitioners when formulating a law? As legislation experts, you should create an accurate legal term to describe the phenomenon and define it clearly. There is no way to judge 'zero/negative-fare' because no benchmarks are available to evaluate whether a GPT is an LFPT or not. ... Shopping and entertainments are two of the six recognized elements of GPTs, how can you simply cut them off entirely from the itinerary? The legislators failed to identify the root problems behind LFPTs. ... They are dealing with problems on an ad hoc basis. They threw the baby out with the bathwater ... I think it is wrong for the government to intervene in the market by administrative means. (Mr ZBQ)

To overcome the above paradoxes, many terms in the CTL2013 need to be re-defined or clarified by adding precise and operable rules:

> The lawmakers did not understand the tourism industry – many of the articles were formulated from a unilateral and biased perspective ... If the CNTA wishes to regulate tour operators through legislation and raise their contractual awareness, the textual representations in the law must be very accurate and indicate operability. (Mr CXJ)

For instance, to facilitate both tour operators' and consumers' judgment of an LFPT, a recommended price range should be provided for a reasonably priced GPT by listing the key costs of its components (e.g. transports, hotels). Moreover, legislators can only obtain a better understanding of tourism practice in reality, and develop more specific rules, by involving more tourism practitioners, consumers, as well as community representatives. Although the draft of the CTL2013 did seek public comments, this study revealed that neither tourism operators nor tourists have been involved or broadly informed in the legislation process.

(b) The supervision and enforcement paradox for relevant administrative departments

Although principles can 'provide flexibility to regulators in the way that they monitor and enforce the regulatory requirements, and can save them from their own lack of foresight in the way detailed rules cannot', regulators may also develop quite conservative interpretations and practices (Black, 2008: 427). The CTL2013 enforcing agencies, such as various TAOs, in this study have adopted such a conservative response because the law did not provide a specific definition or benchmark as to LFPTs, and also failed to delimit each enforcing department's powers and responsibilities. Moreover, Sunstein (1990) has indicated that stringent regulations may cause high regulation cost and inaction by enforcing agencies and the regulated industrial parties because they normally ban cost-benefit balancing or any form of trade-off that is unacceptable to entrepreneurs. The CTL2013 imposed such a stringent control on tour operators, thus arousing their rebellious attitude to the law, leading to an inclination to disregard regulatory controls. This is reflected in the complaints of TA managers that the poor operability of the CTL2013 and the difficulty in joint enforcement caused inaction or weak enforcement by relevant departments (i.e. local TAOs). For example:

> We cannot say local TAOs do not want to strictly regulate TAs' operation following the law. Actually, they do; not only TAOs but also local bureaus of commodity prices (BCP) and Industrial and Commercial Bureaus (ICBs) all do. However, TA management is in a cross-zone of multiple departments – any one of the departments can regulate; but on crucial occasions, anyone may pass the buck to avoid disputes, because the central government has not specified which department should take the main responsibility of supervising TA operation. (Mr CRC)

If local TAOs, ICBs, and BCPs can collaborate to achieve law enforcement jointly, the effectiveness of the CTL2013 could be improved significantly. In addition to the power struggle among government departments, the power relationship between government agencies and enterprises (i.e. TAs) may also lead to collusion. If local TAOs and the TAs that offer LFPTs shield each other by utilizing the loopholes in the CTL2013, it will be challenging for tourists to protect their rights and interests. A real story was provided as follows:

> A guest complained to a local TAO that a screw rod between the two seats on the tour bus punctured his kid's hip and showed the photos.

> The tour bus in the photo looks in a terrible condition – an ancient car with a dirty floor. However, the TAO rejected the complaint, saying that the contract did not specify the condition of the bus but simply indicated that 'a tour bus will be provided throughout the itinerary'. Thus, the consumers should question before signing the contract; but have no right to sue after signing it. (Mr ZBQ)

From this case, we can easily see the favouritism of TAOs towards the tour operators. In conclusion, there are dangerous loopholes in the CTL2013 because the legislators do not fully understand the tourism industry and the clash with market law. Moreover, poor collaboration between various enforcing departments has undermined the effectiveness of the CTL2013. The complaint or prosecution channels for consumers, in reality are still minimal for various reasons.

> Every outbound TA is asked to pay 1.6 million quality margins to the provincial tourism bureau, to compensate the PTTs who suffered loss during the trip and complained after returning. However, as far as I know, in this province, no one tour operator has received complaints because of LFPTs. (Mr ZBQ)

(c) The compliance paradox for regulated bodies (i.e. TAs)

The CTL2013 can facilitate the development of an ethical compliance culture in the tourism industry by providing flexibility to tour operators and allowing them to innovate in the ways in which they comply; but a lack of certainty as to what enforcers will accept as compliance can lead TAs to adopt quite conservative behaviour. In practice, tour operators also take a risk-based approach to compliance, which can hinder the development of self-discipline and business ethics. In the current study, complying with the CTL2013 has become risk management for TAs given the huge market demand. Facing attractive profits, non-compliance became an option for TAs after calculating the gains and losses. Various countermeasures were created: either passive resistance and inaction, or actively seeking countermeasures to get around the law in order to retain profiteering from designated shopping and extra-paid activities. As one TG said: 'LFPTs are still available, even to European destinations; this is an open secret in the industry.' Most practitioners criticized small and medium-sized TAs that disregarded business ethics and reputation for short-term profits. They are viewed as the main culprit in initiating the price war and vicious competition in the GPT market. For example:

> Many designated shops are concealed in the itinerary. TAs will not tell consumers beforehand – TGs will make flexible decisions, such as how many shops to go to and how long to stay at each shop, according to the actual situation. ... Where there are policies, there are countermeasures.

Law enforcement in China has to involve a wide range of departments and various interest groups – the people who have neither legal nor moral sense are not rare. ... On many occasions, laws cannot curb incorrect behaviours - as Dunning said, '100% profit, will make it (capital) ready to trample on all human laws'... Some behaviours need to be restrained by morality. (Mr CRC)

Vast, unjustifiable profit has motivated tour operators to design and promote LFPTs. Some TAs do not hesitate to violate business integrity and regulations for profit-seeking – they have conducted mendacious promotions to mislead or deceive consumers. Such illegal operations must be severely punished and banned. The authors agreed with the interviewees that TAs should improve self-discipline and stop offering LFPTs under the excuse of market demand. If the source of LFPTs is intercepted, consumers will regain their rationality and gradually adapt to a reasonable price. Additionally, feelings of being disempowered and of unfairness were reported as a trigger of the passive resistance of some tour operators to the CTL2013. Almost all practitioners we interviewed claimed that their interests were entirely or partially ignored by the CTL2013, which is over-protecting PTTs' rights and interests.

(d) The last paradox relates to trust

The CTL2013 can lead to the development of an ideal relationship among tourism stakeholders – of responsibility, mutuality, and trust, but these are also the very elements of the regulatory relationships that have to be present for the successful implementation of the law. Without trust – including both PTTs' and tour operators' trust in the legislation and implementation – the CTL2013 will never be enforced.

Paradoxes in consumer mentality

All the interviewees agreed that the actual demand for LFPTs is one of the reasons leading to the ineffectiveness of the CTL2013 in banning LFPTs. Some Chinese PTTs' preference for LFPTs is rooted in their liking for a cheaper price or value for money (Jia, 2006). Chen et al. (2018) found that 'best value for money' is still the predominant criterion of Chinese PTTs when selecting GPTs. However, cheaper price and higher quality is always a paradoxical pair, which normally misleads consumers and causes their dissatisfaction. In addition, pursuing 'best value for money' easily leads to a maximization paradox. A maximization paradox describes a situation where maximizers tend to sacrifice resources (e.g. time and effort) to attain more options, yet such sacrifice ultimately decreases their satisfaction with the chosen option (Dar-Nimrod et al., 2009). Moreover, the provision of LFPTs, in reality may even reduce the satisfaction of normal-price GPT consumers, because once they become aware of the existence of a cheaper option, they may question the utility of their rational choice. In the current study, almost all interviewees reported a strong dislike for LFPTs, but over 60% of them acknowledged its reasonable existence – because of market demand and industry profits. Whether consumers should be blamed for joining LFPTs was controversial among the respondents. Some respondents felt that LFPT participants should be responsible for the prevalence of illegal LFPTs and should bear the ill consequences of their own purchase decision. They assumed that LFPT consumers were knowingly purchasing such unreasonably priced GPTs, and they thus advocated that consumers with malicious intention should not be protected but should be punished by the law. Some LFPT participants did know the consumption traps but still chose these tours for the sake of cheapness and greed or expected a fluke (i.e. consumption by other group members might exempt themselves from the designated shopping). Some even purposefully join LFPTs so that they can request compensation through complaints to TAOs:

> Some tourists go to travel because of the LFPTs, although they do not want to consume at designated shops. Many disputes between LFPT participants and TGs are caused by mismatched expectations toward each other ... Consumers are generally viewed as a disadvantaged group; therefore, if unsatisfied PTTs complain or sue the TA for organising LFPTs, the TA is normally penalised for paying cash or other types of economic compensations. Some tourists have tasted the sweetness, thus constantly attend LFPTs. Perhaps we can say, to a certain degree, the law has stimulated malicious participation. (Mr WZS)

In contrast, the respondents who had participated (intentionally or unintentionally) in LFPTs demonstrated a more tolerant understanding of, and sympathy towards, LFPT participants. These respondents believed that many LFPT consumers might be seduced, cheated, or misled by TAs because it is almost impossible for inexperienced tourists to find out the actual cost of a GPT. No benchmark is available to facilitate their judgment on whether the price of a GPT is reasonable or not. When consumers are surrounded by LFPTs, they may assume such low prices are normal. It is not wrong for consumers to pursue lower prices or better value for money, especially for those who are sensitive to price. In a developing country like China, legislators should consider the affordability of mass tourists. No one, even the government, has the right to deprive economically disadvantaged groups of travel opportunities. As a low-income respondent (#7, Mr LZ, Table 5.1, PTT section) said, the LFPT is a very cost-effective option for the thrifty elderly who are not willing to spend too much money on travel, because most must-see attractions have been included in the itinerary even though the sightseeing time may be less than the time spent in designated shops. More than half the tourists we interviewed could correctly judge whether a GPT given by the interviewer was unreasonably priced or not, indicating that travel experience and the accessibility of relevant information are the key determinants of PTTs' rational judgement. Therefore, the Tourism Law should empower consumers to obtain the right to know – tour operators should improve pricing transparency, mass media and TAOs should input more effort into educating consumers so that PTTs can avoid LFPTs after knowing the actual cost of a GPT. As for those tourists with a gambling mentality or who have malicious consumption motives, penalties should be applied.

Another paradox in the mentality of PTTs is their high dependence on governments psychologically whilst lacking knowledge and trust in law enforcement at the same time. The tourists we interviewed could not distinguish the different duties/responsibilities of various government departments, whilst some knew little about the government structure even. They also demonstrated poor knowledge of the legislative process of the CTL2013. In their discourse, legislators were equated with the central government; different levels of governments and various administrative departments (e.g. CNTA and TAOs) were ambiguously called 'governments'. Furthermore, they believed the government has the responsibility and capability to regulate the chaotic tourism market, but they lack confidence in the lower-level government agencies due to the long-lasting bureaucratic culture (Burns, 1983). For example, the interviewed tourists all believed that the CTL2013 granted them the right to sue. However, in reality, 95% of the respondents had never complained officially, even if they had unpleasant/unsatisfactory experiences with GPTs; and over 60% of them apparently knew to which departments they could appeal – local TAOs or ICBs, National and local consumers associations, as well as the TA that sold them the GPT. The main reasons of giving up the right of petition include: some feel that the complaint procedure is tiring and troubled, and more importantly, no one will seriously handle it; some believe complaint is totally useless because the TA may shield their employees (i.e. TGs), or local TAOs/ICBs may take sides with the TA and reject PTTs' complaints using various excuses. For instance, one respondent had complained about a TG who arbitrarily reduced the number of sightseeing attractions to the consumer hotlines, but they rejected her complaint because she could not provide audio-visual evidence. Some other consumers do not wish the dispute and complaint to ruin their mood of travel. To summarize, PTTs commonly lack trust in law enforcement departments and personnel, and thus feel powerless. The rule requires the plaintiff to provide evidence in civil litigation, which has also hindered many PTTs' complaints or litigation. If the complaint procedure can be effectively simplified, and PTTs can be exempted from providing evidence in certain circumstances, they are more likely to make use of the legal weapons available to protect their own interest and rights, thereby realizing their legal empowerment.

The paradox of moral self-regulation

> Four out of the six TGs in this study expressed their condemnation of the peers who work exclusively for money and lack professional ethics or a sense of honour towards this career: Some TGs had become accomplices of TAs in forcing tourists to purchase, and some TGs performed like deceivers. (Mr CRC)

However, both TA managers and TGs we interviewed showed an understanding of the helplessness of TGs who were involved in LFPTs, because they clearly know the crux of the problem lies in the unreasonable income structure of TGs: 'Many TGs have to undertake LFPTs for making a living.'

Therefore, they often struggle with the dilemma of obeying professional morality and making money (Xu & McGehee, 2017). The majority of the interviewed tourists also expressed their sympathy for TGs, agreeing that being a TG is a tough, challenging job with relatively low income.

As Xu and McGehee (2017: 1106) stated, TGs of LFPTs have to face a moral and ethical dilemma regarding cheating and forcing PTTs to purchase while striving for a commission to 'cover the pitfalls and gain a profit without garnering serious complaints from the tourists'. Similarly, the TGs in this study wish to establish a harmonious relationship with their customers, and thus do not want to benefit themselves at the expense of tourists. TGs describe themselves as scapegoats of the zero-fare tour mode, yet they also enjoy the freedom and profit it offers. This can be explained by the paradox of moral self-regulation – people incline to achieve an internal balance between the desire to act morally and the dislike of paying costs associated with doing well (Sachdeva et al., 2009). TGs generally recognize their collective moral identity has been threatened, and they thus wish to regain their lost moral self-worth by engaging in altruistic/prosocial behaviours – which is called 'moral cleansing' (West & Zhong, 2015). On the contrary, the interviewed TA managers mostly felt they were being licensed to refrain from obeying the CTL2013 because the TAs they were working for had established a reputation as a moral business without LFPTs – this response can be called 'moral licensing' (Sachdeva et al., 2009). Or, some TA managers justified their provision of LFPTs as less illegal by describing the LFPT market as a 'grey-area' or by blaming the ineffectiveness of the CTL2013 on its ambiguous terms, in order to shirk TAs' responsibility (Shalvi et al., 2015). Moral-cleansing and moral-licensing can work together as part of the moral self-regulation process (Sachdeva et al., 2009).

The different attitudes of TGs and TA managers (toward LFPTs and the CTL2013) are predetermined by their different status and associated powers. Ap and Wong (2001) stated that as long as commissions remain the primary source of TGs' income, TGs will remain at the mercy of tour operators and, at times, their unethical business practices. Therefore, the task of reforming the income structure of TGs is extremely urgent. All three groups of stakeholders in this study considered TGs to be a disadvantaged group that also needs better protection and further empowerment by the law. Although Article 38 of the CTL2013 requests TAs to sign a labour contract with, and pay remuneration and social insurance to, TGs – which is very close to the best income structure advised by our respondents – not many TAs in reality have implemented this regulation because of the weak bargaining power of TGs. From this perspective, it may be imperative to establish an independent TG association to protect the interests of TGs. The newly issued 'Measures for the Administration of Tour Guides' (2018) does suggest the establishment of TG associations/branches to protect TGs, but advises that TG associations be subject to the regulations of TA associations and local TAOs. In this case, the disadvantaged status of TGs will not be changed substantially, and the aim of the CTL2013 to protect TGs' interests will remain unrealized.

Conclusion

The primary goal of the CTL2013 is to protect PTTs by regulating tour operators. However, the principle-based regulation approach is less effective in regulating China's disordered tourism market than the rules-based approach – the paradoxes and ambiguities it contains have caused difficulties and challenges for implementers. On the other hand, the disempowered regulated GPT providers actively engage in countermeasures to legitimize LFPTs by utilizing the loopholes in the CTL2013 as well as the standard contract, which has directly resulted in the shortlived and limited effectiveness of the law. Moreover, the empowerment process cannot be fully completed by law enforcement because the CTL2013 is only political empowerment, while contemporary mass tourists may need more psychological empowerment.

This study thus unearthed several paradoxes faced by each group of stakeholders, implying that improving law enforcement and protecting both tourists' and tour operators' rights and interests is not an easy task for any group. Although each party has its own priorities and faces unique paradoxes that need to be resolved, our research suggests that overcoming these paradoxes requires all parties to join hands and work collectively. Legal education and publicity should be expanded to arouse the awareness and self-reflection of all stakeholders, and hence to enhance morals and values when managing stakeholders. These practices may establish the basis for collaboration between all stakeholders in improving the enforceability of the Tourism Law. Otherwise, package tours may be gradually driven out of the tourism marketplace and decline in popularity due to non-intervention by all stakeholders.

References

Ap, J. and Wong, K.K.F. (2001) A case study on tour guiding: Professionalism, issues and problems. *Tourism Management* 22 (5), 551–563.

Atherton, T. (1994) Package holidays: Legal aspects. *Tourism Management* 15 (3), 193–199.

Black, J. (2008) Forms and paradoxes of principles-based regulation. *Capital Markets Law Journal* 3 (4), 425–457.

Burns, J.P. (1983) Reforming China's bureaucracy, 1979–82. *Asian Survey* 23 (6), 692–722.

Chen, H., Weiler, B. and Young, M. (2018) Examining service shortfalls: A case study of Chinese group package tours to Australia. *Journal of Vacation Marketing* 24 (4), 371–386.

Chen, N., Masiero, L. and Hsu, C.H.C. (2019) Chinese outbound tourist preferences for all-inclusive group package tours: A latent class choice model. *Journal of Travel Research* 58 (6), 916–931.

Chen, Y., Mak, B. and Guo, Y. (2011) 'Zero-fare' group tours in China: An analytic framework. *Journal of China Tourism Research* 7 (4), 425–444.

Cohen, E. (1972) Toward a sociology of international tourism. *Social Research* 39 (1), 164–182.

Dar-Nimrod, I., Rawn, C.D., Lehman, D.R. and Schwartz, B. (2009) The maximization paradox: The costs of seeking alternatives. *Personality and Individual Differences* 46 (5), 631–635.

Decrop, A. (1999) Triangulation in qualitative tourism research. *Tourism Management* 20 (1), 157–161.

Dwyer, L., King, B. and Prideaux, B. (2007) The effects of restrictive business practices on Australian inbound package tourism. *Asia Pacific Journal of Tourism Research* 12 (1), 47–64.

Elo, S. and Kyngäs, H. (2008) The qualitative content analysis process. *Journal of Advanced Nursing* 62 (1), 107–115.

Fan, N.N. and Xu, Z.T. (2016) The analysis of influences on domestic travel agencies as a result of the implementation of tourism law. *Journal of Shandong Institute of Commerce and Technology* 16 (3), 6–9.

Fu, F.L. and Jue, P.H. (2015) Several issues for discussion in package tour contracts. *Tourism Tribune* 30 (9), 100–110.

Goodpaster, K.E. (1991) Business ethics and stakeholder analysis. *Business Ethics Quarterly* 53–73.

Grant, D. (1996) The package travel regulations 1992: Damp squib or triumph of self-regulation? *Tourism Management* 17 (5), 319–321.

Jennings, G.R. (2005) Interviewing: A focus on qualitative techniques. In B.W. Ritchie, P. Burns and C. Palmer (eds) *Tourism Research Methods: Integrating Theory with Practice* (pp. 99–117). Wallingford: CABI.

Jia, Y.Q. (2006) The evolution mechanism of 'zero-fare' group tours. *Tourism Science* 20 (1), 56–62.

Kim, J., Bojanic, D.C. and Warnick, R.B. (2009) Price bundling and travel product pricing practices used by online channels of distribution. *Journal of Travel Research* 47 (4), 403–412.

Laplume, A.O., Sonpar, K. and Litz, R.A. (2008) Stakeholder theory: Reviewing a theory that moves us. *Journal of Management* 34 (6), 1152–1189.

Luo, Z.H., Xiao, L.B. and Zhong, L.R. (2017) Comparative study of salary structure of tour guide before and after the implementation of tourism law. *Journal of Jixi University* 17 (2), 54–59.

Ma, E., Qu, C., Hsiao, A. and Jin, X. (2015) Impacts of China tourism law on Chinese outbound travelers and stakeholders: An exploratory discussion. *Journal of China Tourism Research* 11 (3), 229–237.

Prideaux, B., King, B., Dwyer, L. and Hobson, P. (2006) The hidden costs of cheap group tours: A case study of business practices in Australia. In J.S. Chen (ed.) *Advances in Hospitality and Leisure* (Volume 2: pp. 51–71). Bingley: Emerald Group Publishing.

Robson, J. and Robson, I. (1996). From shareholders to stakeholders: Critical issues for tourism marketers. *Tourism Management* 17 (7), 533–540.

Sachdeva, S., Iliev, R. and Medin, D.L. (2009) Sinning saints and saintly sinners: The paradox of moral self-regulation. *Psychological Science* 20 (4), 523–528.

Shalvi, S., Gino, F., Barkan, R. and Ayal, S. (2015) Self-serving justifications: Doing wrong and feeling moral. *Current Directions in Psychological Science* 24 (2), 125–130.

Sunstein, C.R. (1990) Paradoxes of the regulatory state. *The University of Chicago Law Review* 57 (2), 407–441.

Tang, X. (2017) The historical evolution of China's tourism development policies (1949–2013): A quantitative research approach. *Tourism Management* 58, 259–269.

Tse, T.S.M. and Tse, Q.K.T. (2015) The legal aspects of 'zero-fare' tour in shopping tourism: A case of Chinese visitors in Hong Kong. *Journal of China Tourism Research* 11 (3), 297–314.

Wang, W.-F., Chang, Y. and Pearce, P.L. (2018) China's first tourism law: Representations of stakeholders' responses. *Journal of Tourism and Cultural Change* 16 (3), 309–327.

Wang, Y., Weaver, D.B. and Kwek, A. (2015) Beyond the mass tourism stereotype: Power and empowerment in Chinese tour packages. *Journal of Travel Research* 55 (6), 724–737.

West, C. and Zhong, C.-B. (2015) Moral cleansing. *Current Opinion in Psychology* 6, 221–225.

Xu, Y. and McGehee, N.G. (2017) Tour guides under zero-fare mode: Evidence from China. *Current Issues in Tourism* 20 (10), 1088–1109.

Yang, H.X. and Tan, W.Y. (2016) Study on the effectiveness of the implementation of the Chinese tourism law from the perspective of tourists - taking Foshan as an example. *Tourism Management Research* 23 (11), 25–28.

Yin, F. and Zhu, Z.F. (2017) Analysis of the tourism law effects based on tour guide perceptions. *Journal of Hunan Administration Institute* 6, 90–93.

Zhang, Q.H., Yan, Y.Q. and Li, Y. (2009) Understanding the mechanism behind the zero-commission Chinese outbound package tours: Evidence from case studies. *International Journal of Contemporary Hospitality Management* 21 (6), 734–751.

6 Cross-cultural Encounter: Sustaining Racial Prejudice or Prompting Reflection?

Man Tat Cheng

Introduction

Within the perspective that tourism is seen as an instrument to promote peace and celebrate the diversity of humanity, its assumption is that an inter-cultural (including races, ethnicities or socioeconomic shaping) encounter brings about mutual understanding; thus it reduces prejudice and conflict (D'Amore, 1988). This goodwill has also been captured in sustainability language. The UNWTO (World Tourism Organisation, 2015) translates the United Nations' Sustainable Development Goals from the perspective of tourism, in which its elaboration of Goal 16 is relevant to the optimistic view of inter-cultural encounter, as it reads: 'as tourism revolves around billions of encounters between people of diverse cultural backgrounds, the sector can foster multicultural and inter-faith tolerance and understanding...' (Tourism for SDGs, n.d.). The UNWTO idealistically places the tourism sector in the position of a cultural broker, who bears a responsibility to reduce conflict, mediate an encounter or produce change (Jezewski, 1990). Referral to tourism as a singular 'sector' indeed simplifies the multiple chains or layers of tourism suppliers, as well as the broader economic structure within which tourism businesses operate. This study illustrates the circumstances in which the socio-economic structure of the sector and cultural prejudice held by visitors, has conditioned cross-cultural encounters. It reveals the paradox that tourism manifests the space in which the structural and representational view of another culture is reproduced, yet tourism also offers an opportunity for reflexivity, so that such an established view can be challenged.

Our case illustrating this paradox is based in London, a city popular for educational trips by overseas youngsters. Employing an ethnographic method, I volunteered as a helper for a study tour company, travelling with a group of students from the Peoples' Republic of China (China hereafter), in an attempt to observe how they behaved and narrated their experience in the classroom and staying with their host family. In the first part of the chapter the empirical work will be presented, including the context of the researcher's presence in the field, a description of the production of the study tour, and the observed exchanges between the Chinese students and their host families. The following section explores the particular context of Chinese racism, which serves as a reference when we try to understand the behaviour of Chinese young people coming into contact with people from other cultural backgrounds in the UK. The discussion section contextualises the observations in the fieldwork with reference to certain theoretical frameworks, while, at the same time, the paradox will be articulated.

Positionality

It was a 3-week journey travelling with 15 students and 2 adults coming from China. The educational tour operator in London listened to my research purpose. The owner not only kindly offered me a volunteering role in one of his company's group tours, he also answered my questions on the market structure and business practice of the sector. To explain my positionality in the field, I was perceived as an 'elder brother' who studied in the UK and came here as a helper. Although I did tell the students the main purpose of my presence was research, my 'elder brother' position vis-à-vis the students and their innocence (owing to their age) resulted in their not associating themselves with being research subjects. Therefore, my first priority was to ensure their safety and wellbeing. I was vigilant regarding my interactions with the students, so that the conversations and interviews were voluntary, casual and natural, and did not interrupt their learning and travelling experience. At the start of the journey many kept addressing me as 'teacher'. I rejected this and requested that they call me by my first name, with the aim of removing any formality. The participants were assigned to five host families, with consideration being given to their gender, age and body size as grouping criteria. I will detail the diverse experiences of the students, starting from the worst and ending with the best. My gender identity put me in a better position to relate to the male teenagers. I also visited the host families who were hosting two groups of male students. The next section discloses how the business environment and the practices of the educational tour sector have conditioned learners' experience and the institutionalised racism issues observed in Chinese students' attitudes.

Production of the Educational Tour

The London-based firm worked with a large study tour company headquartered in Beijing. To describe the commercial relationship between the two firms, the latter was the customer of the former, as the

Beijing firm recruited Chinese participants through many agencies based in different regions in China. It was expected that the London-based firm would be responsible for local supplies such as providing local transport, classrooms, English language teachers and host families. However, the Beijing firm bypassed their business partner, renting a college in north west London and sourcing a host family agency through its own effort. This was believed to be a way to maximise the profit for the Beijing firm, but the trade-off was that it has little control over the management or negotiation of the service for the entire programme. An observation on this speculative behaviour was that the host family agency did not choose individual hosts whose homes were close to the college in north west London (as ideally the homes would have been). As a result, many participants (having to travel from south London) had to spend more than one hour getting to the school.

The homestay business was competitive. According to one host I talked with, each host family received £20 (US$27) per day per student for an overnight stay, which included breakfast, a packed lunch, and a dinner. It is for this reason that 3-4 students were grouped together to stay with one host family, sharing two bunk beds in a bedroom. I was told by some students that they were given cheap ready-meals. Unfortunately, many of those doing the homestay business are 'black' communities. 'Black' here in the Chinese context could be Jamaican, Indian or another non-white group. Suffice to say that Chinese participants could conflate the host's racial identity with their bad experience rooted in liberal market economy. Talking with the owner of the tour operator, I learnt that he felt frustrated by the situation whereby negative word-of-mouth comments of the participants staying with a 'black' family reached the ears of their parents:

> ... but there are some problems that cannot be solved in China. The problem of racism is extremely serious. Living in the UK we ourselves [Chinese] are sometimes being discriminated against, while they [Chinese participants and their families] look down on Black people, not wanting to live with black host families. Also, last year [2013] one of the tours signed a contract which specifies that students only lived with White host families, but not Black families. In China you can sign a contract like that! Customers taking this contract can ask for compensation...that is to say, because we sent them [participants] to a non-White host family, they asked for compensation. (Owner of the tour operator)

The contract mentioned here was signed between the Beijing partner and the participants. This 'racist contract' explained my observation in the field. Many student groups had already stayed with their host family before they started the English lesson the next morning in the college. There was an 'emergency meeting' held by some tour leaders and the owner of the London company. Complaints were made by students that they had to stay with 'black' host families. One worry was that it would be disastrous if complaints went to the ears of any host families. The conclusion of that meeting was that tour leaders had to ensure that students kept their racist views away from 'black' families.

Cross-cultural Encounters

Worst experience

This group, known as 'Big Girl' group, was comprised of the four eldest students, who were between 15 and 16 years old. I heard about their frustrations from time to time. I did not hear complaints about the fact that they had to spend more than two hours travelling between the accommodation and the college every day. These girls were disappointed in the way they were treated by the Jamaican host family. Four of them were put together in a bedroom. Although the host promised to provide one more room for them upon their arrival, at the end they were still squeezed into one bedroom. Unfortunately, the room was also shared by the host's daughter, who stored her toys there, which meant that even less space was available for the visiting students. The girls felt particularly irritated because of the low quality of food provided by the host. They complained that they did not have enough food to eat for dinner. They had to share a pizza that cost £1 (US$1.4). Sometimes dinner was two packs of twin burgers that cost £1 each, shared by the four of them.

It was the last day of the journey, and host families were required to accompany the students to the college, according to the terms and conditions they signed with the tour operator. However, the host of Big Girl group did not fulfil this requirement. Her friend, who was also hosting three younger girls living close to her, eventually accompanied two girl groups to the college. One member of the Big Girl group told me that the whole experience was upsetting and insisted that she would make a complaint upon returning to China. She asked me if I knew how much each host had earned during their stay, because she thought that her host family was so mean to her and her friends. Before that day, the girls did not associate what they had found unacceptable about the host family with their ethnicity or race, but finally one moaned: 'I never discriminated against Black people, but this time I could not help thinking in this way... .'

Bad experience

The 'Small Boy' group was made up of four teenagers aged 12–13 years. Earlier I found out that some students complained about being assigned a 'black' host. This included two members in this group who had stayed with an Indian family. They were moved to a Caucasian host the same day, joining the two female teachers who were the only two being put with the white family. They complained that the Indian family was not 'normal', because there was only an old woman and her daughter in the family. They explained to me that it was a commercial guest house, with three rooms on the ground floor and six on the first floor. They saw other Chinese students living temporarily in the same place. Food was again an issue. I heard similar stories from one of the boys who remained staying with the Indian family. I was shocked by the immediate move that was organised for the two participants. Had a racist contract been signed? This speculation gains some support, perhaps, from the fact that I found that the Beijing company had to bear the extra fee for the Caucasian host. The other two participants were less disappointed, which was partly due to different expectations, and perhaps partly due to the lower fee they paid to the tour operator.

Fantastic experience

The bad experience lasted for only one day and the rest of the home-stay experience for these two juveniles was fantastic, so it was felt by the two teachers, who often shared with me how lucky they were to have stayed with their family. The Caucasian Englishman and his wife, a Thai woman, had two children. The students this time had a chance to experience a 'normal family', as they told me that the two small children of the family always asked them questions. In the second week, I heard from the teachers that they were thinking about a birthday gift for the host's birthday, and I saw that one student bought a mug for him in the National Gallery. Unlike other host families, this host took his guests to Windsor castle one Sunday. It was the last week of the journey and, as I had developed friendships with the group, and given that these teachers and students had had an enjoyable experience with the host family, I asked if I could visit them. The host kindly asked me, through the teachers, if I would like to have dinner with them, and thus I had a chance to spend an evening with these four participants and the family.

Upon entering his house together with the participants, one boy immediately asked the host if he could go 'swimming' in the garden. I realised that it had become a routine that he had been playing in a 6–8 feet wide paddling pool, which was still not spacious enough for this small 12-year-old boy. The Englishman showed me his garden and birds, and we had a casual conversation. He, the teachers, and I were chatting after dinner in the lounge, and soon were joined by his wife, who had returned home from work. I kept in touch with the teachers after the summer tour (so did they and the host family), when the teachers told me that they received an email from the host. It appeared that their pleasant home-stay experience was not constrained by the relatively large group number. The host's personality was the defining factor. During the whole trip, none of them had discussed the racial identity of the host family.

Warm experience

Finally, it was the 'Big Boy' group's experience in which we see their substantial encounter with a Jamaican family. The four youngsters, who were older or bigger in size, stayed with a Jamaican family. The parents and their son also lived in south London. The Big Boy group did not complain about the time wasted in travelling. In this case, it seems that Louisa, the female host, in her 50s, was the reason for the warm experience felt by the boys. Many times, the students remarked how delicious the food cooked by Louisa was. 'She always said, "Oh thank you darling, good boy", one student said, imitating Louisa's tones and gestures. Another expressed how she left him with a very good impression of British people. These students even created a group entitled 'Louisa' on their Chinese social networking site. Considering their good relationship with the host, and their gender and older age, again I proposed a visit to their family. The two teachers were happy for me to do so. The Big Boy group was also excited to have me visit them. Louisa kindly accepted the request of the students and I visited the family after afternoon sightseeing. It was already around 5 pm before a short bus journey took us to the home, when the students said they needed to go to McDonald's. I realised that this was part of a daily routine for them, because the food provided by the host was not enough. When we got to the host's home, the students were immediately engaged with their gadgets for games and comics, and I had a casual chat with Louisa and had my perception confirmed that she was a very nice person. Louisa worked as an ambulance driver. It was a Christian family, and I observed that the lounge was a specific place dedicated to worship.

I was chatting with Louisa while she was preparing the dinner. In the meantime, I heard that one student was playing piano and Louisa expressed to me he was very talented as a beginner. I said that I was not intending to stay for dinner, since I was not confident that the teenagers had been able to inform Louisa about my visit in sufficient time. Louisa tried to persuade me to eat something, while I stressed that I had already had an early dinner. She was heating up four packs of cottage pie, which were in ready-meal packages. I also observed that she was about to cook a bag of pasta, as she said that a large portion of dinner was needed since the youngsters had a very good appetite. I saw the students eating on their own, then I realised that the students and the family did not eat together, because other family members came home late, as they told me. Dinner was served in relatively large portions, but the quantities were not sufficient for the boys. They felt embarrassed to ask for more from Louisa, which explained why they filled themselves up in advance with fast food. After dinner, I was moving between the two rooms shared by the students. They shared with me some more experiences of staying with the host family: how strong the son of Louisa was; an instance when one of the boys walked into an unlocked door of the bathroom twice while others (fortunately a male family member) were taking a shower; and how they spent time playing games with Louisa's nieces on one of the Sundays. The youngsters made some racist jokes about the appearance of these young people, such as 'among Black people they are sort of pretty', in which they emphasised that this was not discrimination, but an 'aesthetics judgement'. I felt that the teenagers genuinely enjoyed the kindness of and their interaction with the family, especially Louisa, irrespective of their hosts' ethnic/racial origins or the size of the food portions.

Chinese Racial Discourse

The market economy of the study tour and the homestay business, as I observed in the field, reveals that all parties were trying hard to squeeze their profit from the business at the expense of the participants' wellbeing. Under these conditions, the quality of cross-cultural contact is largely shaped by the economic factor; unfortunately, the racial identity of the host, as perceived by the students, takes its toll on such quality.

One limitation of this study is that that part of the structure of the homestay business relevant to socioeconomic and ethnic factors is not explored. The parameters within which this chapter works to make sense (Weick, 1993) of some of the paradoxically negative and positive consequences observed in the field. How can we interpret these Chinese young people's racist acts? The question of whether racism exists in China is perhaps easy to answer when we compare it to a normative standard in the west. However, what is more meaningful, perhaps, is to understand the contextual formation of the racist view that exists in China, in which such a view is not an ethical issue. This section is an attempt to establish the perspectives drawn mainly from one historian's scholarly efforts as well as critical reflections on the quotidian spaces in China where 'race matters' could be observed.

Racial discourse

How to define racism? This is a perennial ontological question that a wide range of answers could never compromise on, apart from a general view that it has negative connotations. Today we see people across cultures and geographies behave divergently towards the racism issue. The *Guardian* columnist, Carmen Fishwick (2014), born in the UK and of Vietnamese origin, regards her frequent experience in London of being complimented on her English accent as 'casual racism'. In China, the Lunar New Year variety show in 2018, watched by up to 800 million people worldwide, included a performance that was intended to celebrate relations with Africa. A Chinese actress appeared in blackface make-up, with her breasts and buttocks enlarged and a basket of fruit on her head, using an 'African accent' to speak and sing in Mandarin on the stage (*Independent*, 2018). In both examples, those who praised Fishwick and the variety show producer could be called racist; however, they could also legitimately respond by claiming they are not, drawing from their own ethnocentric frame (Rattansi, 2007). These frames are interwoven with socio-economic, cultural and geopolitical factors. The proposition of this study towards racism is therefore marked by ambivalence and contradiction. That is to say, people with different ethnocentric frames might not share the property or degree of racist behaviour. More precisely, a people might not be aware of how they may offend another racial group, owing to the particular structure that is shaping them. Exploring this structure is the analytical approach of this study. The concept of racial discourse is used as an analytical tool.

Doane's (2006) work exemplifies the power of exploring racial discourse in the United States. It examines the language and expressions used in debates in the public arena, such as in university campuses, police reports or radio programmes. It also traces the implications of different discursive framings of racism with respect to a contestation of political power and economic resources. Doane has not referenced which school of thought of *discourse* is employed in her analysis, apart from giving this definition of *racial discourse* – 'the collective text and talk of society with respect to issues of race' (Doane, 2006: 256). This is akin to Hall's 'common-sense' conception of discourse as 'a way of representing – a particular kind of knowledge about a topic' (Hall, 1996: 201). Treating racism within the field of discourse is not without objection. Garcia (2001) criticises from an ethical perspective that progressive action to tackle racism is displaced by retrospective discussion of the subject. More importantly, a discursive perspective on racism, referred to by the author as inherently immoral in any situation, plays down individuals' responsibility for acting in a racist manner, i.e. their motives and intentions. This argument is very reasonable in its own right, but there appears to be a methodological problem: it is not always easy or possible to read a person's inner world. Again, the ethnocentric factor of a people mediates the implication of a perceived racist act. Ignorance, misunderstanding or prejudice shall not be an excuse for demeaning another group; however, agency has its limits. A racist could be themself be a victim. The value of analysing racial discourse is *first of all* in its attempt to understand the netting of racism issues.

The discourse of superior Han Chinese

Establishing race relations in contemporary China, it is important to elaborate the unique ethno-racial identity of the Chinese and how 'black' racial groups are being identified and represented. It features race relations across an historical timeline, with the aim not to claim that the dominant Chinese group *inherits* a particular worldview on racial issues, but to demonstrate the coherence of racial discourse in various temporal and spatial contexts. The paramount context of Chinese race relations is the high homogeneity that categorises Chinese-ness. This identity is both external and internal. Roger Yonchien Tsien, who was awarded the Nobel Prize in chemistry in 2008, is widely regarded as a Chinese American. Tsien was born to biologically Chinese parents in the United States but he does not write or speak Chinese. His phenotype or ancestral heritage makes him Chinese, as observed from how he is presented by the Nobel Prize official website (Nobel Foundation, 2008). That Tsien received his PhD from the University of Cambridge is perceived as an honour by the university. The university's webpage constructs the tie between China and the university as follows:

> Cambridge alumni include the poet Xu Zhimo, the author Louis Cha, the Nobel Laureate Roger Tsien, and ...... Today, Chinese students form the largest international group from a single country at Cambridge. (University of Cambridge, 2015: n.p.)

It is expected, presuming that you do not know the appearance of Tsien, that you could have formed a mental representation of his phenotype – which is not typical of the Uyghurs or Kazakhs from Xinjiang, the Tibetans or the Mongols, or any other races residing in the PRC. Readers would most probably associate Tsien in a category called the Han Chinese, which is claimed to be a homogenous group in China and the container of common ancestral origins and shared cultures (Dikötter, 2006). When Tsien was asked by Chinese journalists in Stockholm about whether he was Chinese, used Chinese language, and what the significance of his achievement was to Chinese scientists (Liberty Times Net, 2008), he answered:

> I was born and raised in the United States, and I can not speak a lot of Chinese. So, I don't think I'm a Chinese scientist, rather I am an American. (Lin, 2013: n.p.)

This example demonstrates the common identification of a Chinese person by racial taxonomies, emphasising physical attributes and 'common blood'. This framing also applies to how Han Chinese view themselves. Chinese people typically use 'yellow skin', 'yellow race', 'black eyes', 'black hair', as symbolic markers to differentiate themselves from others such as Caucasians or blacks. These representations are important in the construction of the racial nationalism of the contemporary Chinese (Cheng, 2011; Dikötter, 1994; Sautman, 1994).

The term *Han* was named after the powerful Han Dynasty, which is famous for the great military power that expanded the territory of the dynasty. Han people had a sense of cultural superiority. Han also refers to the traditional culture, which is equivalent to the virtuous Confucian values rooted in ancient Chinese Confucian classics. To sum up this Sinocentric worldview, Chinese people are a representation of Han, which serves as a cultural, ethnic and racial indentification of such a population. Dutch historian and sinologist Dikötter states:

> 'Chineseness' is seen to be primarily a matter of biological descent, physical appearance and congenital inheritance. Cultural features, such as 'Chinese civilization' or 'Confucianism,' are thought to be the product of that imagined biological group. (Dikötter, 1994: 404)

Historical hierarchical race relations

Ethno-racial Sinocentrism has a strong bearing on hierarchical race relations in China. I have selected the period of Maoist China, a time when more cross-cultural contact between Chinese and African communities began to take place and were documented. In the early 1960s, black scholar Hevi (1963) detailed how African students experienced abuse when they studied in China. They were viewed as unhygienic by Chinese students. African students' Chinese girlfriends were arrested by the authorities. European and African students were treated unequally. The former had RMB 150 monthly allowance while the latter had only RMB 80. Liu's (2013) analysis of Chinese state-classified official documents verifies Hevi's account. It furthermore reveals the suffering of African students in Beijing, reporting that they were made a spectacle of as though in a zoo (MOFAA, 1961a).

Propaganda circulated in Maoist China serves as further historical material with which to analyse race relations. In China, the 1950s and 1960s saw a bullish mood of anti-imperialism and anti-capitalism. Mao proclaimed that the Chinese people were leading the fight against white imperialism. African countries were allied by China in what was known as a Third World coalition. Mao used the racial discourse to represent how the 'black' and the 'yellow' were both suffering from the imperialist power of the West, so the Africans and the Chinese had to unite. Mao once befriended Africans ironically when he met African visitors in 1960:

> [Westerners] say we Chinese are useless, we coloured people are useless, we are dirty, and we are not elegant. Our race seems to be the same as you Africans. (Ministry of Foreign Affairs Archives [MOFAA], 1960)

In late 1963, Premier Zhou Enlai visited African countries, starting a new chapter of Sino-African relations. Zhou announced a range of strategic measures to establish a closer relationship with African countries. This responded to the racist behaviour of the Chinese towards Africans. For example, Chinese people were reported to have surrounded black people and shouted at them, while Chinese women refused to shake hands and kept their children away (Shanghai City Archives, 1956). Ghana's ambassador in Beijing was greeted with yells of 'his face is like bark' and 'his head resembles a pig's' (MOFAA, 1996b). Two propaganda posters reflect the official state policies that tried to alter the negative attitudes of Chinese citizens towards African people. Figure 6.1 shows Africans reading the *Little Red Book*, which contains quotations from Chairman Mao, published from 1964 to 1976. My interpretation of the meaning of the poster, in relation to race relations, is this. African people, like Chinese, also embrace Chairman Mao and we belong to the same class, i.e. the proletariat. Now there are no *differences* between the two groups. Figure 6.2 depicts not only the friendship of both peoples, symbolised by the African child extending its arms in welcome of the Chinese nurse, but 'raises' the 'inferior' status of African representation that they can be 'wise enough' to be a doctor (i.e. the man dressed in a

Figure 6.1 Chairman Mao is the great liberator of the world's revolutionary people
Note: Printed in Chinese: 毛主席是世界革命人民的大救星.
Translation of printing: Chairman Mao is the great liberator of the world's revolutionary people.
Designer: Unknown.
Publisher: Shanghai renmin meishu chubanshe (Shanghai People Art Publication).
Date: April 1968.
Source: Chineseposter.net (2005). Part of the IISH/Stefan R. Landsberger/Private Collection.

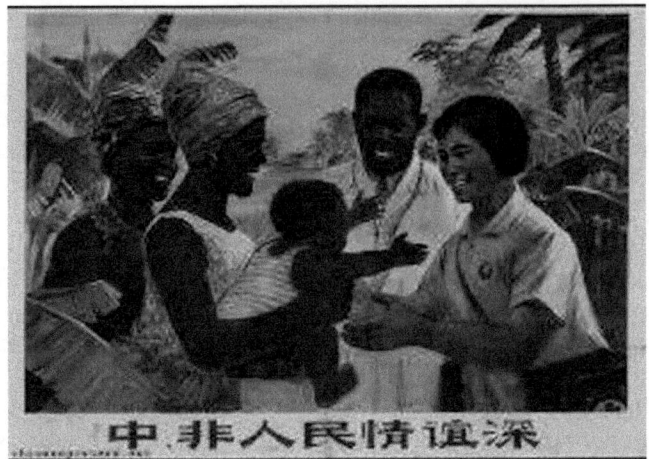

Figure 6.2 The feelings of friendship between the peoples of China and Africa are deep
Note: Printed in Chinese: 中, 非人民情谊深.
Translation of printing: The feelings of friendship between the peoples of China and Africa are deep.
Designer: Changzhoushi gongnongbing meishu chuangzuo xuexiban gonggao (Workers and Farmers Group Art Creation Class in Changzhou City.
Publisher: Shanghai renmin meishu chubanshe (Shanghai People Art Publication).
Date: April 1972.
Source: Chineseposter.net (2005). Part of the IISH/Stefan R. Landsberger/Private Collection.

white coat). The symbolic message is that now it is time to stop looking down on black people.

Contemporary racial discourse

Moving to the present, I have chosen one case to demonstrate racial discourse. It is the story of a bi-racial girl, Lou Jing, who appeared on a TV show in 2009 and stirred heated debate about her national/racial identities (Frazier & Zhang, 2014). The show was a platform for new faces to impress the audience with their charm, wit, voice and dancing skills. Her skin colour, rather than her performance ability, was more important for turning her into a popular star. When she first appeared on the stage Lou Jing was presented by the host as: 'Our chocolate girl', 'black pearl' and 'Halle Berry of the East', followed by hip hop music that accompanied her rapping a self-introduction. The producers predicted the background of Lou Jing would provide a focus for the show, offering substantial airtime to interview her about her past as well as her mother's relationship with an African-American father, who left China without knowing of her pregnancy. Social media and technologies have provided an interactive platform for the public to comment upon and construct Lou Jing's identity. I learnt about Lou Jing from Frazier and Zhang (2014) and further explored the latest blog messages about Lou Jing.

The content of the blogosphere ranges from racist to anti-racist comments. It also contains contestation on the meaning of authentic Chinese identity, and topics that revolve around nationalism and modernisation. Extreme racist comments include 'black chimpanzee', 'black devil' or 'black slave race'. Lou Jing's mother was 'truly lamentable', 'truly both low and deplorable' and had 'little self-respect'. These comments indicate not only the taint of unmarried pregnancy – in traditional Chinese values pregnancy outside marriage in China is considered immoral behaviour – but also her relationship with a black person, and the view that she gave birth to a 'pathetic biracial child' (Fauna, 2009). The comparison of black and white is also a focus. An online user (NetEase, 2009) argues that '…white represents jade and white embraces the meaning of health, beauty and cleanness…'. This post concludes that if a child is bi-racial, i.e. European-Chinese, there would be no problem. Lou Jing revealed her discomfort at being unaccepted by Chinese society: 'it was okay if I didn't speak. When I talked, people would start to discuss it' (Key, 2009). It was Lou Jing's insistence about her Chinese identity that 'upset' some bloggers: 'our nation shall never accept the existence of this kind of shame……she has made all Chinese lose face'. Fortunately, there were many commentators who came to the rescue of Lou Jing and her mother, condemning the racist comments, such as those who 'strongly despise people who are racial discriminators!!! China is a tolerant nation'. Some comments were intertwined with nationalism and modernisation discourses.

To sum up, the core property of racial discourse in China is the superiority-inferiority dichotomy. Skin colour is an effective marker in attributing an inferior status to black people. The abuse of the bi-racial girl indicates that even her home-grown Chinese cultural belonging does not earn her equal status to a Chinese in general. In other words, one-drop policy (even though she is half Chinese, people focus on her blackness) applies to her external identification. Finally, from the instances that I have drawn, it is observed that a debate on black people could invoke a reflection of the status of Chinese group identity. Sometimes this reflection is bound with nationalistic sentiment, interwoven with the legacy of the historical struggle against the imperialist invasion.

Discussion

In presenting the analysis of Chinese racial discourse the aim is not to essentialise the Chinese teenagers' worldview. Rather, it is an attempt to demonstrate that their particular life experience pertaining to race relations is a normality, instead of an ethical issue that needs to be dealt with. It helps us to understand the context of a cross-cultural encounter, which is more effective than just condemning racist behaviour. Weick's (1993) sensemaking offers a rich framework with which to organise our response and action to everyday circumstances. Its first step involves noticing and comprehending a circumstance, retrospection of the plausible stories behind such circumstance (Weick et al., 2005). In the context of tourist studies, McCabe (2005: 86) comments that understanding of tourist experience focuses on existential knowledge of individuals, overlooking 'the importance of the wider social discourse of tourism in shaping and defining individuals' versions of their experience'. Tourists are not entering into a foreign and unfamiliar place empty minded: they carry tools of discourse available in their home (Moore, 2002).

The most disturbing finding of the study is that there existed a 'racist contact', signed between the Beijing tour operator and their customers. The firm was obliged to offer a 'white' host family for the participants. It was observed that some students complained about being housed with a 'Black' host family. A plausible explanation to these incidents of explicit racism is that they were not as great a taboo as expressing a dislike of black people in China. Historically there were African visitors in China who were subjected to abhorrent racist chanting in the 1960s; and nowadays in the public space (TV show) and in the virtual space (the blogosphere), a bi-racial girl and her mother were abused. The expression and circulation of this attitude towards another racial group has become a normality and a socialisation, which is not known to be against the law in China. The responses given by the participants offer a clue to understanding this structure. When the female students were moaning about how they were mistreated by their host, one of them claimed that she herself never discriminated against black people, but what she had experienced was just unacceptable. Another young man made a racist comment – 'among black people they are sort of pretty' – but at the same time distanced himself from discrimination, claiming rather that it was an 'aesthetics judgement'. A plausible explanation for these responses is that explicit racism prevailed in China. Compared to the racism they have observed in daily life in China, the insensitive comment as shown above is far from bracketed in the racism category. It is also these particular moments that invoke the circumstances of race relations in their home country. The study tour offered an opportunity for the students to undergo a process of reflexivity, as they were made to compare their personal experience in the UK with race relations in China, and subsequently evaluate and articulate their responses to a given situation.

The notion of reflexivity has been richly discussed in different disciplines. Foucault's (1969) archaeology of knowledge is one classic analytical method that prompts human reflexivity. The prevalent racial discourse in China in the Foucauldian sense is accumulation of knowledge. 'He [Foucault] reminds us of the influence of anonymous rules that we unknowingly or knowingly apply, that these rules are beckoned and reinforced by institutions and are strengthened by putative practices, which bring to the fore the possibility of misrepresentation...' (Bartilet, 2014: 32–33). Reflexivity takes place when the young people experience the warm hospitality and friendship offered by a racial group who in their home country are brutally abused. Reflexivity is triggered when they consciously reject the chance to be labelled racist. Archer's (2003) internal conversation explains this internal mechanism when they tried to maintain with themselves, redefining their beliefs and attitudes. In tourism research, tourist reflexivity could occur after the trip through narrating their experiences to other people, which is a way to create a sense of self (Desforges, 2000) and even transformation of self (Noy, 2004). Then, the memories of cross-cultural encounters experienced by the Chinese participants will remain to be brought back in particular times and spaces against the various backdrops of their life situations.

Conclusion

A cross-cultural encounter is always an ethical issue; therefore the UNWTO aims to achieve tolerance and understanding in the encounter. But it appears problematic when the tourism sector is assigned the responsibility (Tourism for SDGs, n.d.). This empirical study has demonstrated a common inherent characteristic of the problem that some tourism businesses might not share the same UNWTO's Sustainability Goal in the free market economy. The organisation of a cross-cultural encounter is governed by the economic motive, which is often intertwined heavily with the socio-ethnic factor in some places. In this case, the unfavourable living conditions offered by some non-White host families have given a negative cross-cultural encounter experience to the Chinese youngsters, who could have connected the host behaviour with their racial identity. And unfortunately, this has matched with the omnipresent racial discourse they have gathered in quotidian exchanges in China. From this perspective, the tourism sector is responsible for reinforcing the conditions leading to racism. Therefore, scholars who are advocates of peace tourism argue that we need to critique the capitalist model of peace-through-economic development (Blanchard & Higgins-Desbiolles, 2013). The tourism sector is far from being the actor that initiates a change of the political economy of the sector (Wohlmuther & Wintersteiner, 2014). Paradoxically, without this opportunity of cultural exchanges, optimists of peace tourism will miss an evidence to show the positive side of a cross-cultural encounter. The peoples have had more knowledge and understanding of each other, and made possible a new friendship. These echo the contact hypothesis (Allport,

1954), which believes sufficient contact between groups could reduce prejudice and stereotype. The analysis of personal reflexivity in this chapter has offered another dimension of positive outcome observed in the encounters.

References

Allport, G.W. (1954) *The Nature of Prejudice*. Oxford: Addison-Wesley.
Archer, M.S. (2003) *Structure, Agency and the Internal Conversation*. Cambridge: Cambridge University Press.
Bartilet, J.L. (2014) Foucault, discourse, and the call for reflexivity. *Filocracia* 1 (1), 24–34.
Billig, M. (1995) *Banal Nationalism* (Vol. 1). London: Sage.
Blanchard, L.A. and Higgins-Desbiolles, F. (2013) *Peace through Tourism: Promoting Human Security through International Citizenship*. Abingdon: Routledge.
Cheng, Y. (2011) From campus racism to cyber racism: Discourse of race and Chinese nationalism. *The China Quarterly* 207, 561–579.
D'Amore, L.J. (1988) Tourism: A vital force for peace. *Tourism Management* 9 (2), 151–154.
Desforges, L. (2000) Traveling the world: Identity and travel biography. *Annals of Tourism Research* 27 (4), 926–945.
Dikötter, F. (1992) *The Discourse of Race in Modern China*. Stanford, CA: Stanford University Press.
Dikötter, F. (1994) Racial identities in China: Context and meaning. *The China Quarterly* 138, 404–412.
Dikötter, F. (2006) *Things Modern: Material Culture and Everyday Life in China*. London: Hurst.
Doane, A. (2006) What is racism? Racial discourse and racial politics. *Critical Sociology* 32 (2–3), 255–274.
The Economist (2014) Coming to a beach near you: How the growing Chinese middle class is changing the global tourism industry. Magazine article, accessed 12 May 2015.
Fauna. (2009) Shanghai 'black girl' Lou Jing abused by racist netizens. Magazine article, accessed 1 September 2009. https://www.chinasmack.com/shanghai-black-girl-lou-jing-racist-chinese-netizens.
Fishwick, C. (2014) The BBC not acting on Clarkson's racist comment shows its disregard for us. Columnist letter, accessed 29 July 2014. https://www.theguardian.com/commentisfree/2014/jul/29/bbc-clarkson-racist-comment-east-asian-minority.
Foucault, M. (1969) *The Archaeology of Knowledge*. (Translated by A. M. Sheridan Smith.) London & New York: Routledge.
Frazier, R.T. and Zhang, L. (2014) Ethnic identity and racial contestation in cyberspace: Deconstructing the Chineseness of Lou Jing. *China Information* 28 (2), 237–258.
Garcia, J.L.A. (2001) Racism and racial discourse. *The Philosophical Forum* 32 (2), 125–145.
Hall, S. (1996) The West and the rest: Discourse and power. In S. Hall, D. Held, D. Hubert and K. Thompson (eds) *Modernity: An Introduction to Modern Societies* (Volume 1: pp. 184–224). Oxford: Blackwell.
Hevi, E.J. (1963) *An African Student in China*. London: Pall Hall Press.
The Independent. (2018) Chinese New Year 2018: TV celebration gala sparks outrage as 'racist' performance includes blackface sketch. Newspaper article, accessed 16 February 2018. https://www.independent.co.uk/news/world/asia/chinese-new-year-2018-blackface-sketch-tv-gala-china-celebration-a8213471.html.
Jezewski, M.A. (1990) Culture brokering in migrant farmworker health care. *Western Journal of Nursing Research* 12 (4), 497–513.
Key. (2009) NetEase interview with Shanghai black girl Lou Jing. Magazine article, accessed 15 September 2009. http://www.chinahush.com/2009/09/15/netease-interview-with-shanghai-black-girl-lou-jing/.

Liberty Times Net. (2008) De jiang le Zhongguo qiang zhan guang. Newspaper article, accessed 20 May 2015. http://news.ltn.com.tw/news/world/paper/249385.
Lin, R. (2013) China's thirst for Nobel. Newspaper article, accessed 15 June 2015. http://www.sino-us.com/10/Chinas-hope-prospects-of-winning-Nobel-prize.html.
Liu, P.H. (2013) Petty annoyances? Revisiting John Emmanuel Hevi's an *African Student in China* after 50 Years. *China: An International Journal* 11 (1), 131–145.
McCabe, S. (2005) 'Who is a tourist?' A critical review. *Tourist Studies* 5 (1), 85–106.
MOFAA. (1960) Mao Zedong meeting African visitors. Minutes. #102-00036-03.
MOFAA. (1961a) Situation briefing. Foreign Affairs Committee of Beijing City, in situation of students from Cameroon and other countries in China. Minutes. #108-00695-01.
MOFAA. (1996b) Teaching people how to correctly treat black African visitors. Minutes. #117-01299-01.
Moore, K. (2002) The discursive tourist. In G.M.S. Dann (ed.) *The Tourist as a Metaphor of the Social World* (pp. 1–17). Wallingford: CAB International.
NetEase. (2009) NetEase's exclusive interview with Lou Jing: I am a native Chinese. Post comments, accessed 14 September 2009. http://comment.news.163.com/news2_bbs/5J4GFANS00012Q9L.html.
Nobel Foundation (2008) Roger Y. Tsien: Biographical. Biography, accessed 11 November 2015. http://www.nobelprize.org/nobel_prizes/chemistry/laureates/2008/tsien-bio.html.
Noy, C. (2004) This trip really changed me: Backpackers. *Annals of Tourism Research* 31 (1), 78–102.
Rattansi, A. (2007) *Racism: A Very Short Introduction*. New York, NY: Oxford University Press.
Rotary International. (2014) A guide for home families. Organisational mission, accessed 25 March 2015. https://www.rotary.org/en/document/66101.
Sautman, B. (1994) Anti-black racism in post-Mao China. *The China Quarterly* 138, 413–437.
Shanghai City Archives (1956) To pay attention to black visitors. Shanghai city's committee of foreign services. Memorandum. #B255-2-90-91.
Tourism for SDGs. (n.d.) SDG 16 – peace justice and strong institutions. UNWTO resources. Accessed 22 March 2019. http://tourism4sdgs.org/sdg-16-peace-justice-institutions/.
Tucker, H. and Lynch, P. (2005) Host-guest dating: The potential of improving the customer experience through host-guest psychographic matching. *Journal of Quality Assurance in Hospitality & Tourism* 5 (2–4), 11–32.
University of Cambridge (2015) Cambridge established the UK's first Professorship of Chinese in 1888, and the University has educated many leading Chinese scholars and Sinologists. University website, accessed 11 November 2015. https://www.cam.ac.uk/global-cambridge/regional-focus/china.
Weick, K. (1993) The collapse of sensemaking in organizations: The Mann Gulch disaster. *Administrative Science Quarterly* 38 (4), 628–652.
Weick, K.E., Sutcliffe, K.M. and Obstfeld, D. (2005) Organizing and the process of sensemaking. *Organization Science* 16 (4), 409–421.
Wohlmuther, C. and Wintersteiner, W. (2014) *International Handbook on Tourism and Peace*. Austria: Drava.
World Tourism Organization. (2015) *Tourism and the Sustainable Development Goals*. Madrid: UNWTO.

7 Contemporary Polemics of Chinese Outbound Tourism to Europe: Paradoxes, Inconsistencies and Contradictions

Rose de Vrieze-McBean

The People's Republic of China (PRC) has been undergoing a profound transformation in the past few decades, and four fundamental philosophies – Capitalism, Confucianism, Communism and Consumerism – are emerging to suggest that the republic is seeking to modify its image on the world stage. Yet, despite the appearance that China has developed into a seemingly open country, which is in part grounded on the 'Reform and Opening-up' business model instigated almost four decades ago, the People's Republic of China is fundamentally still a socialist state. This implies that its governing body exercises political authority from the centre, and indeed, such practices are evident in the country's outbound tourism policies. Kotler *et al.* (2006: 132) claim that 'The political environment is made up of laws, government agencies and pressure groups that influence and limit the activities of various organisations and individuals in society.' For example, the role that the Ministry of Culture and Tourism (MCT), formerly known as the China National Tourism Administration (CNTA), plays in Chinese outbound tourism, appears to be a political one. The MCT not only supervises the quality of services, but it also manages the legal rights of tourists and operators, and regulates tourism operations and the services of tourist enterprises and tourism practitioners (OECD, 2016).

According to the Chairman of the CNTA, since the implementation of the Reform and Opening-up Policy, China's tourism development has skyrocketed: the nation has become one of the world's major tourist destinations and it continues to refine its tourism product. Powered by the innovative technological revolution and digital information dissemination, China's modernization is producing significant paradoxes not only regarding people's lifestyle, but also in its outbound tourism industry (CNTA, 2018). This is evident in the fact that the CNTA website is predominantly run by the government. Likewise, it is a measure applied by the said Administration to execute its responsibility of information disclosure. Besides, it issues travel policies and guidelines, while functioning as a vital interface to showcase the latest developments of its tourism product when communicating government information on tourism. Furthermore, the organization uses various platforms to inform tourism businesses, the tourists themselves and the general public. In so doing, it endeavours to extend its influence and to gain both national and international recognition, ultimately constructing understanding within general society about the importance of the tourism industry (CNTA, 2018). However, in its efforts to organize its outbound tourism products and services, paradoxes, inconsistencies and contradictions have emerged. This chapter focuses on these, predominantly with reference to current Chinese outbound tourism to Europe.

At the heart of these paradoxes, inconsistencies and contradictions lies the recent announcement that the CNTA and the Ministry of Culture have united to form the Ministry of Culture and Tourism (MCT) (Arlt, 2016b). According to the State Councillor, Wang Yong, this unity of government bodies 'is aimed at coordinating the development of cultural and tourism industries, enhancing the country's soft power and cultural influence, and promoting cultural exchanges internationally'. The concept of 'soft power' here is likened to Joseph Nye's reference to soft power as '... the ability to shape the preferences of others through appeal and attraction' (Nye, 2004). Not only is soft power grounded on non-coercion: it is based on cultural, political and foreign policies.

This is why the MCT embarks on constructing rules, establishing surveys, safeguarding cultural capital and controlling the generic tourism market. Simultaneously, cultural exchange with foreign countries is being stimulated (Arlt, 2016b). This is not only significant for China's outbound tourism industry but also for China's image enrichment through its soft power influence. The fact that tourism has been elevated to ministerial status will have a significant impact on both domestic and international tourism (*China Daily*, 2018). Moreover, it is gathering the Republic's support of its outbound tourism industry, despite the huge amount of expenditure occurring abroad by Chinese travellers. Hence, tourism exhibits four fundamental inconsistencies, particularly within European societies: firstly, as a product of Capitalism; secondly as a product of Confucianism; thirdly as a product of Communism and finally as a product of Consumerism.

Chinese Outbound Tourism as a Product of Capitalism

The first inconsistency when inspecting Chinese outbound tourism to Europe is that tourism is a product of capitalism. Here, a distinction can be made between the 'Chinese model of capitalism' and the 'Western model' of capitalism. Whereas in the West, capitalism is depicted as a liberal market economy with private enterprise at its centre, the Chinese model is controlled by the state (Tselichtchev, 2012). However, there are visible shifts in the current economic system in China and the country is displaying similarities with the Western model of capitalism. According to Tselichtchev (2012), they both involve shrewd business players, and both have pecuniary objectives. Additionally, they exhibit equivalent capitalistic and institutional practices such as market transactions, open competition, private enterprises resulting in profits and so on. Yet there are differences, which particularly lie in norms and values as well as cultural rituals (Zhao, 2015). Within the Chinese model of capitalism, traditional Chinese cultural forms are evident in social actions such as 'face', 'relationships' and 'human sentiment'. Furthermore, their goal for pursuing 'private' or material interests is often covert as they are obliged to observe moral ethical codes of Confucianism (Zhao, 2015), which is a crucial element of the Chinese mode of capitalism.

Chinese outbound tourism is playing an ever-increasing role in capitalism. The Communist Party of China (CPC) has affirmed that the republic need not forsake its obligation to its ultimate ideology, so long as it acknowledges that capitalism is a necessary part of its progression. Bell (2008) argues that 'The CPC need not abandon the commitment to communism as the long-term goal so long as it recognises that poor countries must go through capitalism on the way.' As in the capital mode of production, workers are treated as mere tools in the productive process and technology is put to use for the purpose of enriching a small minority of capitalists (Bell, 2008). In China, for example, travelling to Europe has become a 'must do' event, especially for the younger Chinese traveller. Paradoxically, this was once considered a privilege for the affluent Chinese national only. These days, however, it has become a commonplace occurrence for any Chinese to visit Europe; some even on multiple occasions. In fact, in the first quarter of 2018, the number of Chinese arrivals to Europe doubled compared to the same period the previous year. According to the MCT, in 2018 an estimated 71 million Chinese nationals travelled abroad for the first time (TTR Weekly, 2018). This was the result of a major promotion campaign enacted by Chinese and European governments to boost the relationship between Europe and China. As part of a strategic plan to further enhance Sino-Euro relations, 2018 was designated 'China-Europe Tourist Year', and Chinese tourists enjoyed more convenient access in terms of visas and flights, which heightened trips to Europe (European Travel Commission (ETC), 2018). Thus, the desire to travel to Europe could be seen as a product of capitalism seeping through China's outbound tourism.

Historically speaking, the Chinese Communist Party, which had been the ruling party in China since 1949, was steered by the economic revolution of Deng Xiaoping and prolonged by Jiang Zemin. It shifted its

Administration (CNTA) from a government-controlled planned economy to a socialist market economy (Qian & Wu, 2000). This implied that, instead of the state determining production and pricing, consumers' demand for goods and their production costs govern the quantity and price, while maintaining the economic market model of supply and demand. In spite of this shift in model, the government sustains its communist control on the economy by applying laws and regulations in its self-established market economy. Furthermore, the socialist market economy demonstrates a spread of capitalism in China, where government regulations do not support a completely free market. Hence, China is not merely emulating a Western mode of capitalism. Indeed, on the contrary, Milillo contends that, unless the government's interference subsides, Western-style capitalism is unlikely to be fully realized in China (Milillo, 2008).

Chinese outbound tourism, however, is being seen – particularly in foreign destinations such as Europe – both as a product of and a boost to capitalism. For example, David Jobanputra expresses his opinion on travel as expansionism in his travel blog on *Tourdust*. In his article, Jobanputra claims that major Western countries engage in a constant pursuit for virgin lands on which to propagate their surplus capital. He further contends that capitalism has sprouted more wealth than anyone knows what to do with, while simultaneously consuming most of the West's natural resources, such as picturesque landscapes, pristine nature, etc., in which to invest even more capital (Jobanputra, 2018a). Although it is assumed that developed countries send tourists to less developed countries, this is not the case with China, where, as a less advanced country, it sends tourists to advanced countries. In 2016 alone, the aggregate of Chinese direct investment in the European Union was 35 billion euros. This amount, which included all sectors together, was 77% more than the corresponding figure in 2015 (Amato, 2018).

Not only are Chinese tourists seeking out the famous city destinations in Europe, they are also developing a taste for its unfamiliar regions. It is, therefore, not surprising to discover in particular young Chinese visitors in remote areas of Europe, seeking quintessential experiences and engaging in novel events with the local population. Sheep-shearing in Ireland, or enjoying the annual Tour de France cycling competition in Europe, for instance, are no longer considered unusual tourism activities for Chinese tourists.

Moreover, Fulcher (2004) claims that the substantial growth of international tourism is one of the most striking displays of the increasing economic influences among nations, as it spreads capitalist practices to regions of the world that have scarcely been affected by the advancement of capitalism. This dissemination of international tourism is driving consumer spending and is generating a larger demand for food production and transportation, and at the same time supplying the base for souvenir manufacturing and replication of artefacts. Consequently, the income from tourism can intensify currency circulation, leading to the importation of manufactured goods and the establishment of innovative configurations in consumption patterns (Fulcher, 2004). Fulcher further argues that leisure breathes capitalism, because it demands continuous work, long working hours and provides non-work activities, such as leisure for all workers. This proves to be more productive than having work disrupted by taking days off. Leisure in itself is therefore not only capitalism-in-action but it leads to commercialization: workers pay for leisure activities organized by capitalist enterprises, which in turn produces 'mass leisure' in the form of 'mass tourism'. Paradoxically, novel businesses and industries that emerge, serve to exploit and develop the leisure market even further resulting in an enormous source of consumer demand, employment and huge profits. Hence, tourism becomes a product of capitalism (Fulcher, 2004).

Tourism as a Product of Confucianism

The second inconsistency when reviewing Chinese outbound tourism to Europe is that tourism is a product of Confucianism. This belief system is often portrayed as one of social and ethical convictions rather than one of religion. However, Confucianism was established on an early religious paradigm intended to introduce the social values, institutions, and supreme ideals of a conventional Chinese society. The renowned sociologist, Robert Bellah, referred to Confucianism as a 'civil religion' (Bellah, 1975). He reasons that it is the sense of religious identity and common moral understanding centred at the foundation of a society's central institutions. It is also what a Chinese sociologist called a 'diffused religion': its institutions were unlike churches but more like a society, family, school, or state; its priests were not separate liturgical specialists but parents, teachers, and officials. Confucianism was part of the Chinese social fabric and way of life; to Confucians, everyday life was the amphitheatre of religion (Bellah, 1975).

Subconsciously, Confucianism appears to play a meaningful role in the lives of many Chinese tourists as they travel through Europe. They seem to be strongly influenced by the teachings of Confucius (551–479 BCE) and they consequently adhere to values such as 'harmony' and 'respect' for authority. Kwek and Lee (2010) suggest that the belief in harmony is intricately related to themes such as 'respect' for 'authority', 'relationship building' or Guanxi (relationship building) and 'conformity'. Chinese travellers are greatly influenced by the values of Confucianism, placing emphasis on maintaining not only the correct and/or appropriate behaviour but also showing respect to authority. Nonetheless, there are some inconsistencies, especially with regards to Chinese millennial tourists and free independent travellers, who apparently do not seem to adhere to Confucian values, mainly because they do not travel in large groups. Likewise, these tourists consider values such as 'benevolence' and 'conformity' to be 'outdated'.

When travelling in Europe, it is chiefly the older Chinese nationals who travel on package or group tours. Data gathered from holiday travel bookings produced by CTrips and Huayuan International Travel, indicate that 67% of Chinese nationals still elect to travel on package tours (TTR Weekly, 2018). This mode of travel is not uncommon among these travellers for three main reasons. First the language barrier. Older Chinese tourists hardly speak English. Neither are they comfortable using trendy applications (Apps), which could alleviate their language deficiencies. Hence, they prefer to have a guide who not only has local knowledge of the destinations being visited but also speaks either the local language or is able to converse in English to the local population. Second, the issue of safety remains a concern. Despite the numerous measures being taken in some European destinations to keep Chinese tourists safe from criminals, there are frequent reports of Chinese nationals being robbed. Finally, shopping is one of the most common reasons why Chinese tourists visit Europe. Luxury items, in particular, are relatively cheaper compared with their cost back home in China. In addition to this, they are rather sceptical about products bought in China. According to one Chinese millennial, Chinese nationals do not trust certain goods purchased at home because they are often counterfeit or of an inferior quality. Consequently, these inconsistencies are reflected in the behaviour of Chinese tourists when abroad: they travel in large groups based on economic, safety, linguistic and other societal reasons. They spend considerable sums on famous brands and prefer to shop in Europe due to their belief that certain goods are cheaper and more authentic there as opposed to back home.

However, younger Chinese tourists seem to be gradually moving away from Confucian values and are more concerned with the more pressing issue of 'finding themselves' (de Vrieze-McBean, 2019). So, although package tours remain highly popular, the number of Chinese tourists visiting Europe independently is rising sharply. Confucian values such as 'harmony', 'conformity' and 'Guanxi' do not seem to be as relevant to them as they are to their parents. Contrastingly, they seek freedom to choose their own destinations and prefer to do this either alone or with chosen friends. They also pursue interaction with the locals, and seek authentic experiences, which they brag about on their social media platform(s). This is prompting some tourism providers to re-vamp their products to undertake considerable investment and innovation to meet the demands of this new (growing) generation of visitors. Therefore, while most Chinese tourists still demonstrate some Confucian behaviour during their travel experience, young Chinese travellers are, ironically, departing from most of these traditional Confucian values in their pursuit of their own happiness. This is evident in the places they visit and the activities they eventually participate in.

Tourism as a Product of Communism

The third inconsistency when examining Chinese outbound tourism to Europe is that tourism is a product of Communism. Increasingly Communism or Red Tourism is spreading, not only among older Chinese tourists but also among the younger ones. In the PRC, Red Tourism, or *'Pinyin'* in Chinese, is defined as a sub-set of tourism in which Chinese people visit locations with historical significance to Chinese communism 'to rekindle their long-lost sense of class struggle and proletarian principles' (Zhou, 2010). Most of these sites are located in the republic itself: e.g. Shaoshan, the birthplace of Mao Zedong; Jinggangshan, the cradle of the Chinese Communist revolution where Mao Zedong and other leading members of the Communist Party of China founded the first rural base for the revolution in 1927; and Nanjie, in Henan province, a small village where the local residents still live under Maoist ideas and according to commune principles (Zhou, 2010). Red Tourism destinations are also very popular in former communist countries in Europe. Examples are the Czech Republic, which was ruled by the Communist Party of Czechoslovakia (the former Czech Republic); and the birthplaces of Karl Marx (Trier, Germany, 1818) and Friedrich Engels (Wuppertal, Germany, 1820).

Wolfgang Arlt, who writes for *Forbes International*, expressed the view that 'communism is weaving itself into another part of modern Chinese culture: international tourism'. Red Tourism has been a long-established phenomenon, representing more than four billion domestic trips made to sites significant to the history of China's Communist Party (Arlt, 2016a). Within China, people began eagerly supporting Red Tourism to promote the 'national ethos' and socioeconomic development in those areas that were typically rural and poorer than East China; as one Chinese official stated: 'This is a major project that benefits both the Party, the nation and the people, either in the economic, cultural and the political sense' (People's Daily Online, 2005). This online newspaper also stressed; 'It will make people, especially the young people, further consolidate their faith in pursuing the road to socialism with Chinese characteristics and realizing the great rejuvenation of the nation under the leadership of the CPC.' The official went on to say: 'visiting these sights, will ensure valuable assets of national character-building and independence among young people' (People's Daily Online, 2005).

However, Red Tourism is no longer limited to China. Research conducted by the Chinese Outbound Tourism Research Institute (COTRI) on behalf of Engels House in Wuppertal, Germany, indicated that, from 2011, besides Russia, Chinese tour operators had already been organizing 'red'-themed tours to Germany, the United Kingdom, and former Yugoslavia. These tours highlighted places with links to figures like Marx, Engels, Lenin and Tito (Arlt, 2016b). Although the number of the travellers to these international destinations has grown significantly, the composition of the groups is generally senior citizens. Nevertheless, there is also a segment of the market comprised of multi-generational family-groups. Young Chinese who accompany their parents (usually a 'dream' of the parents) on such a mission, visit certain countries in order to experience the impact that Communism has had on their parents' young development. Concurrently, Red Tourism could provide an educational experience for these young Chinese in that they could have a better understanding of their parents' cultural heritage (Arlt, 2016b).

International tourism is also expanding its ground in the form of new destinations that commemorate the history of the Chinese Communist Party abroad. This not only exhibits Red Tourism outside China, but also gives it 'Chinese characteristics' (Arlt, 2016b). For instance, the building that accommodated the wealthy Chinese gentlemen who helped to bring Chinese students to France to study and work after the collapse of the Qing Dynasty in 1912, has been turned into a memorial hall. This was the home of notables such as Mao Zedong and Deng Xiaoping. At this memorial hall, courses in Mandarin, calligraphy and the art of the Chinese tea ceremony are offered to the local population. It also attracts Chinese tourists to the city and these visitors are proud to find a statue of Deng Xiaoping, centred in the square named after him (Arlt, 2016b). Arlt concludes that, with an outbound tourism market of roughly €170 billion, there will definitely be more places capitalizing on their Chinese Communist Party heritage (Arlt, 2016b). In line with this, and according to Denton (2014), Red Tourism has lately adopted global dimensions as part of Xi Jinping's One Belt, One Road initiative, and China and Russia have introduced the so-called 'Sino-Russian Red Tourism Large-scale Exchange Activities'. Consequently, Red Tourism has been intensely publicized as a sociopolitical instrument.

Within China, a vast majority of Chinese tourists experience Red Tourism as a communal, participatory, performative and ritualized event (Denton, 2014). Frequently, many Chinese tourists to revolutionary sites are part of tour groups organized by factories, schools, state and private corporations. For example, Denton (2014) talks of one specific plant that sponsored a tour group to Jinggangshan. There they marched along the 'Carrying Grain Path', on which Zhu, De and Mao supposedly chivalrously delivered grains to their soldiers. Donned in Red Army uniforms and caps, these workers wanted to gain a deeper understanding of the 'Jinggangshan spirit'. Thus, Red tourism is incorporated into a company's ideological training and so employees are expected to diligently participate in such exercises. Therefore, company employees sometimes go on such excursions as an essential constituent of their educational experience. Chinese entrepreneurs and businessmen also make use of such training to stimulate team-building, and to create camaraderie in predominantly large corporations (Denton, 2014: 4).

Contrastingly, such rhetoric is hardly demonstrated in Communism (Red) tourism in Europe, where such practices still remain highly controversial. Recently, the city of Trier in Germany celebrated the 200th birthday of Karl Marx and, to mark this occasion, it received a towering bronze statue of Marx as a gift from China's Communist government. Marx, who was a socialist philosopher and one of the founding fathers of communism, was born in this southwestern German town (Speitkamp, 2017a). Paradoxically, this statue, which was created and manufactured in China, was welcomed with mixed feelings. Among the many reasons for discontent, was the massive inflow of tourists who made a pilgrimage to this small town. In addition to this, human rights groups and organizations representing the victims of communism voiced their disapproval of this 'mega-Marx' bronze image, due to the fact that Karl Marx is a controversial figure, whose writings served as the basis for a whole state form that later spun into dictatorships (Speitkamp, 2017a).

Nevertheless, 'The connection with China has, of course, a touristic purpose', according to Speitkamp, a former history professor at Kassel University in Germany (Speitkamp, 2017a). The same is true for the Engels' statue in Wuppertal. Fredrich Engels is the other co-founder of Communism and, like Karl Marx, he is a polemic figure in Germany. That is why these monuments attract throngs of Chinese tourists to Germany, particularly to towns like Trier and Wuppertal. Conversely, in China, Marx and Engels are not only foremost philosophers, they are vigorously studied throughout the nation (Speitkamp, 2017a). Thus, ironically, the increasing demand for Red Tourism is powering an inflow of tourism from China to these European destinations.

Despite the growing demand for Red Tourism, most Chinese millennials are not especially interested in the ideological and revolutionary ideals of the CPC. Like many international tourists, young Chinese tourists travel – even to revolutionary sites, – to have fun, to relax, eat and drink, to see the sites and maybe come away with some unique insights into the destination. Their relationship with the past, if any, is of a bygone nature. Most Red tourists are from urban regions while the sites themselves are mostly to be found in rural areas, so there is spatial distance between the consumer and the consumed (Speitkamp, 2017b). Seeing that the revolution is no longer something to be lived, Red Tourism contributes only to the re-making (Hollinshead, 2009) of the revolution as a 'tradition'; one that the state superficially uses as a historical and sociopolitical platform for generating tourism revenue to drive its socioeconomic plans, as well as to enrich the lives of the rural population (Denton, 2014). Therefore, tourism as a product of Communism demonstrates inconsistencies in that it is powered by the sociopolitical and radical principles of the Chinese Communist Party, while many Chinese tourists in fact just travel to these sites merely to consume tourism products.

Tourism as a Product of Consumerism

The fourth and final inconsistency when scrutinizing Chinese outbound tourism to Europe is that tourism is a product of Consumerism. Consumption has been considered a preeminent post-modern undertaking, and for the first time in history entire societies are engaged in this universal action of consumption (Jobanputra, 2018a). Not only do we buy what we need to subsist, but we are also constantly seeking commodities that we perceive will make us 'happy' or 'fulfilled'. Commodities such as, safety, comfort, beauty, health, learning, leisure, love and status, are fervently pursued. We are inclined to believe that such possessions empower us and that they will provide us with a sense of inclusion and belonging (Jobanputra, 2018b). Shopping and sightseeing hence are two main reasons why Chinese tourists travel to Europe. They are often steered as lambs to the slaughter to gigantic shopping malls, which have dotted the continent over the past decades. These new consumption factories are mostly populated by luxury items and goods that attempt to meet the seemingly insatiable desire of especially the wealthy Chinese visitor, who merely searches for the latest must-haves. This phenomenon can be categorized as the age of consumerism (Jobanputra, 2018a).

In this current age of consumerism, everything is commoditized; and the virulent lucidity of marketing synthesizes it to tourism products: shopping outlets, the new mall in the outback or in the small town, flight packages to Europe, souvenirs, postcards, anything that will provide Chinese tourists with that unique experience. Images of such lived experiences are subsequently posted on their favourite social media page to boost their 'likes' and supply them with that coveted status. With the rapid increase in flights to Europe brought on by the 'opening up' policy of the Chinese government, millions of Chinese tourists are finding their way to the continent. European destinations are also relaxing travel visas to facilitate the new itinerants.

Seeing that China's enormous population has only just begun to travel, their visits will have a significant impact on the world (Lomas, 2017). Despite the recent debate on China's imminent economic slowdown, Chinese consumers are expected to spend increasingly more, particularly on travelling. According to a report by the McKinsey Global Institute, by 2020 there will be approximately 400 million 'mainstream consumers – with disposable household incomes of roughly US$ 34,000 per annum. It is this cohort of consumers that will "rock" the world' (McKinsey Global Institute, 2019). Additionally, Euromonitor International estimates that China will contribute more than any other country to global consumer expenditure growth (Euromonitor International, 2017a).

The more affluent that China's consumers become, the more dynamic the landscape becomes: price is no longer the only dominant factor. These Chinese consumers increasingly focus more on brand, quality and the status attached to the product purchased (McKinsey & Company, 2019). Consequently, tourist destinations wishing to attract Chinese tourists need to bear this in mind when attempting to lure these consumers. As previously mentioned, Chinese consumers do not trust domestic products as there is a danger that these may be fake or counterfeit. Hence, they give preference to purchasing quality brands when travelling in Europe.

Similarly, health and safety are of great concern, especially after numerous scandals in China affecting a number of consumer products. The result of this is that Chinese travellers to Europe, for example, are eager to buy European products, which they consider more trustworthy, authentic and relatively cheap. They trust products bought in Europe more, including make-up and personal accessories.

Another factor that has been well documented is the growing number of people over 60 in China. Euromonitor International (2017a) claims that China will have 345 million people over the age of 60 by 2030. This is noteworthy because with such an aging population, demand for healthcare services and products will soar. Though the state is taking measures to alleviate this problem by increasing healthcare spending, hospitals are said to be far behind other developing economies. Consequently, there has been a surge in health tourism; and 'consumers aren't waiting anymore… when people get cancer, they're getting on planes to Singapore and London' (Towson & Woetzel, 2017).

It is not just the older generation that is becoming more health conscious and are, therefore, spending more on their health: 'Many Chinese consumers have an innate distrust of processed foods and that, combined with the recent food scandals have driven many consumers to seek out what they consider to be safer and healthier foods' (Euromonitor International, 2017b). Furthermore, this is creating new health-trends. Yoga, for instance, is one such trend. Another is cycling. The bicycle, which was once considered a cheap mode of transport/travel, is now considered a 'healthy way' to go. So, Chinese consumers are increasingly opting for yoga and cycling holidays in Europe (Chinese tourist informant).

Meanwhile, better-educated heads of households are driving rising consumption in some of the world's most dynamic markets (Euromonitor International, 2017b). Goldman Sachs Global Investment Research (2015: 33) published a report titled *The Asian Consumer* in which a depiction is made of China's fun-seekers and their spending habits. Understandably, Chinese millennials seem to value 'fun' more than their predecessors did. These youngsters visit the cinema and eat out on a regular basis – particularly at shopping malls offering unique experiences (Goldman Sachs, 2015: 34). In addition to this, they are travelling the world and changing destinations as they go. This is significant in that their interests and motivations are somewhat different from their parents. Whereas the latter travel in large groups and are inclined to favour the bigger cities while in Europe, Chinese millennials are often seen in smaller cities and towns. They also participate in (local) sporting events. These days, it not surprising to see Chinese tourists both visiting large festivals and concerts in Amsterdam and tasting wine in the South of France.

Conclusion

In conclusion, present-day discourses of Chinese outbound tourism to Europe consist of paradoxes and inconsistencies in that they are grounded on four distinct and often contradictory tourism pillars: Capitalism, Confucianism, Communism and Consumerism. First, leisure and recreation were made possible by the perception of capitalism. In China, the socialist market economy propagated capitalism there, even though this was a sharp contradiction of what the state had intended. In fact, the Chinese government does not support a completely free market economy, and thus the notion of capitalism is not to be compared to that of the West. Second, many Chinese tourists who visit Europe are said to display Confucian values during their travels. Harmony (which includes respect for authority), Guanxi (relationship building) and conformity (adherence to authority) are three distinctive Confucian values displayed by many Chinese tourists during their travels. Nevertheless, many Chinese millennials are increasingly departing from these values. They contend that notions such as conformity and benevolence are 'old-fashioned', hence they seek unique and local experience while travelling through Europe.

The third tourism pillar is Communism. Tourism as a communist product exhibits inconsistency, considering it is powered by the ideological and revolutionary ideas of the Chinese Communist Party even though many Chinese tourists travel to 'Red' sites just to have fun, relax and to have a good time. This final point links to suggestions that consumerism has become one of the main reasons why Chinese tourists visit Europe. A desire to shop for luxury items, to visit museums and to acquire unique and quintessential experiences while in Europe, appears to be the strongest reason for visiting the continent. These consumers are expected to spend more on travel in the coming years. Concurrently, Chinese tourists are visiting Europe to purchase luxury goods and services, seeing that they no longer seem to trust buying such goods at home. The Chinese population is also aging, prompting new challenges in healthcare. While the government is bent on making provisions for its senior citizens, many of them are seeking help abroad, where there are ample destinations eager to meet this 'health tourism' demand.

Finally, the four fundamental philosophies discussed in this chapter: capitalism, Confucianism, communism and consumerism could suggest that the republic is seeking to modify its image on the world stage. They operate paradoxically when combined together to influence

Chinese outbound tourism to Europe. In particular, these pillars serve to underpin the characteristics of these visitors and at the same time highlight the impact of Chinese outbound tourism on European destinations, both now and likely in the years to come.

References

Amato, R. (2018) *Tourism in Focus: China Tourism 2018*. Directorate-General for Internal Market, Industry, Entrepreneurship and SMEs. Virtual Tourism Observatory. Brussels: European Commission. https://ec.europa.eu/growth/tools-databases.vto/ [Retrieved December 29, 2019].

Arlt, W.G. (2016a) Countries want Chinese tourists – just not all in the same place. *Forbes International*, 22 September 2016. https://www.forbes.com/sites/profdrwolfganggarlt/2016/09/22/countries-want-chinese-tourists-just-not-all-in-the-same-place-585952955c0d [Retrieved December 28, 2019].

Arlt, W.G. (2016b) How communism influences where Chinese tourists travel. In Forbes Asia # *Wanderlust*, 25 September 2018. https://www.forbes.com/sites/profdrwolfganggarlt/2016/09/25/how-communism-influences-where-Chinese-tourists-travel/ [Retrieved December 2019].

Bell, D. (2008) A new openness to the world. *The Guardian*, 25 August 2008. https://www.theguardian.com/commentisfree/2008/aug/25/olympics2008.china1 [Retrieved December 28, 2019].

Bellah, R.N. (1975) *The Broken Covenant: American Civil Religion in a Time of Trial*. New York, NY: Seabury Press.

China National Tourism Administration. (CNTA) (2018) March 10. China National Tourism Administration Information Centre. webmaster@webmastercnta.gov.cn.

Confucius (551–479 B.C.) Analects: Confucius quotes, brainy quotes. https://www/brainyquotes.com/authors/confucius [Retrieved April 2018].

Denton, K.A. (2014) *Exhibiting the Past: Historical Memory and the Politics of Museums in Postsocialist China* (chapter 10). Honolulu, HI: University of Hawaii Press.

De Vrieze-McBean, E.R. (2019) A multidimensional inquiry into Chinese outbound tourism to western Europe: The visitation of Chinese millennial students to the Netherlands. PhD thesis, University of Bedfordshire. Ipskamp Printing Proefschriften.net.

Euromonitor International (2017a) China's impact in global economy and consumers in 2017. Euromonitor Research, 17 March 2017. https://blog.euromonitor.com/chinas-impact-on-global-economy-and-consumers-in-2017/ [Retrieved December 30, 2019].

Euromonitor International (2017b) Future prospects of Chinese outbound travel for the lodging industry. March 2017. https://www.euromonitor.com/future-prospects-of-chinese-outbound-travel-for-the-lodging-industry/report [Retrieved January 3, 2020].

European Travel Commission (ETC) (2017) The European Travel Commission welcomes the designation of 2018 as the EU-China Year for Tourism. https://ec.europa.eu/growth/content/2018-eu-china-tourism-year-0_en [Retrieved December 29, 2019].

Fulcher, J. (2004) *Capitalism: A Very Short Introduction*. Oxford: Oxford University Press.

Goldman Sachs (2015) *The Asian Consumer: The Chinese Tourist Boom*. Report by Goldman Sachs, November 2015. https://www.goldmansachs.com/insights/pages/macroeconomic-insights-folder/chinese-tourist-boom/report.pdf.

Hollinshead, K. (2009) Tourism and the social production of culture and place: Critical conceptualisations on the projection of location. *Tourism Analysis* 13 (5/6), 639–660.

Jobanputra, D. (2018a) Travel as imperialism. Tourdust blog. https://www.tourdust.com/blog/posts/travel-imperialism [Retrieved December 28, 2019].

Jobanputra, D. (2018b) Travel: The ultimate must-have possession? Tourdust blog. https://www.tourdust.com/blog/posts/is-travel-anything-more-than-consumerism-in-another-land [Retrieved December 28, 2019].

Kotler, P., Bowen, J. and Makens, J. (2006) *Marketing for Hospitality and Tourism* (4th edn). Upper Saddle River, NJ: Kotler, Bowen & Makens.

Kwek, A. and Lee, Y.-S. (2010) Chinese tourists and Confucianism. *Asia Pacific Journal of Tourism Research* 15 (2), 129–141. https://doi.org/10.1080/10941661003629946 [Retrieved December 2019].

Lomas, M. (2017) Chinese consumers will change the global economy. *The Diplomat*, 6 April 2017. https://the diplomat.com/2017/04/chinese-consumers-will-change-the-global-economy/ [Retrieved December 2019].

McKinsey & Company (2019) *China Luxury Report 2019: How Young Chinese Consumers Are Reshaping Global Luxury*. McKinsey Greater China's Apparel, Fashion and Luxury Group. https://www.mckinsey.com/~/media/mckinsey/featured%20insights/china/how%20young%20chinese%20consumers%20are%20reshaping%20global%20luxury/mckinsey-china-luxury-report-2019-how-young-chinese-consumers-are-reshaping-global-luxury.ashx [Retrieved March 10, 2020].

McKinsey Global Institute (2019) *China and the World: Inside the Dynamics of a Changing Relationship*. By Jonathan Woetzel, Jeongmin Seong, Nick Leung, Joe Ngai, James Manyika, Anu Madgavkar, Susan Lund, and Andrey Mironenko. 1 July 2019. https://www.mckinsey.com/~/media/mckinsey/featured%20insights/china/china%20and%20the%20world%20inside%20the%20dynamics%20of%20a%20changing%20relationship/mgi-china-and-the-world-full-report-june-2019-vf.ashx. [Retrieved December 2019].

Milillo, N. (2008) Capitalism 2008: A look at the Beijing Olympics and their effect on China. http://www.neumann.edu/about/publications/NeumannBusinessReview/journal/Review_SP06/pdf/china.olympics.pdf. [Retrieved December 2019].

Nye, J. (2004) *Power in the Global Information Age: From Realism to Globalism*. Abingdon: Routledge.

Organisation for Economic Corporation and Development (2016) *OECD's Tourism Trends and Policies 2016: Policy Highlights*. https://www.oecd.org/industry/tourism/Tourism2016-Highlights_Web_Final.pdf [Retrieved December 2019].

People's Daily Online (2005) China boosts 'red tourism' in revolutionary bases. http://english.peopledaily.com.cn/200502/22/eng20050222_174213.html [Retrieved December 30, 2019].

Qian, Y. and Wu, J. (2000) China's transition to a market economy: How far across the river? Working Paper 69, King Center on Global Development, Stanford University.

Speitkamp, W. (2017a) Karl Marx and Friedrich Engels monuments in Germany: A present from China. Deutsche Welle, 15 March 2017.https://www.dw.com/en/why-a-marx-monument-is-still-controversial-in-germany/a-37948036.

Speitkamp, W. (2017b) Why a Marx monument is still controversial in Germany. Deutsche Welle, 15 March 2017. https://www.dw.com/en/why-a-marx-monument-is-still-controversial-in-germany/a-37948036 [Retrieved September 23, 2020].

Towson, J. and Woetzel, J. (2017) *The One Hour China Book: Two Peking University Professors Explain All of China Business in Six Short Stories*. Columbus, OH: Towson Group LLC.

Tselichtchev, I. (2012) *China Versus the West: The Global Shift of the 21st Century*. Singapore: Wiley.

TTR Weekly (2018) Chinese tourists flock to Europe. 29 November 2018. https://www.ttrweekly.com/site/2018/11/chinese-tourists-flock-to-europe/ [Retrieved December 30, 2019].

Wang, K. (2018) *China Daily*. Tourism, culture merger seen as mutually beneficial. http://www.chinadaily.com.cn/a/201803/14/WS5aa879aea3106e7dcc1417eb.html [Retrieved 14 March 2018].

Zhao, W. (2015) What is capitalism with Chinese characteristics? Perspective on state, market, and society. Colloque International. Recherche & Regulation 2015. Conference. RR-2015_Zhao_88568. Pdf. P1-15.

Zhou, Q. (2010) National symposium on Red tourism kicks off in Zingtan University. Hunan Government.

8 International Tourism Academia: A Paradoxical Challenge

Vincent Platenkamp

'What is being stated here is not true'

Introduction

In this chapter, I would like to try to design an escape route out of the paradoxical trap of one's own perspective. I shall dwell upon the nature of a paradox and how, for example, Bertrand Russell tried to deal with its consequences by distinguishing language levels. I will explore the paradoxical situation in the international tourism academy. If the dominant character of this academy is Anglo-Saxon, then it is not diverse. If it is diverse, then English should not be the dominant language. How, then, could one develop a diverse international tourism academy? I will then describe some of the suffocating effects of the concept of a paradigm on the different islands inhabited by our tourism tribes. These paradigms do not lead us out of our overwhelming paradoxes in the international tourism academy, but rather the contrary. At the same time, I will compare this situation to the journey of Odysseus to Ithaca. Finally, I would like to propose the situation of Ithaca as a metaphor for Kumasi (in Ghana), and to elaborate on how this metaphorical action could help us to escape from our killing and paradigmatic paradox, by using Russell's solution to paradoxes.

Paradoxes and Russell's Types

In the history of philosophy, paradoxes have always played an important role. Consider the folk paradox: Which came first, the chicken or the egg? As we all know, every chicken comes from an egg, but every egg comes from a chicken. Many answers seem to be right. People such as Darwin responded to this paradox in their own way. Aristotle once said that *each* chicken comes from an egg and *each* egg from a chicken. Nobody can give a general answer to the problem. Darwin said that there are only chickens for a finite amount of time. Therefore, in the beginning eggs must have preceded chickens or vice versa. One must conclude: this discussion does not seem to be very satisfying. Achilles and the Tortoise is a famous paradox, attributed to of Zeno of Elea (Aster, 1968). Zeno convinces his audience that when the tortoise is given a small head start, the fleet-footed Achilles will never overtake the tortoise. As Lewis Carroll's Alice said, children know that something is wrong but cannot put their finger on it.

An interesting contemporary paradox is the paradox of freedom, as Popper (1945: note to chapter 7) has put forward. Unlimited freedom, he states, is logically impossible because then you have to give freedom to those who want to destroy freedom. On the basis of this paradox, there will always be a discussion on how much freedom civilians are willing to concede to the authorities in order to protect their freedom. Or, as the Dutch scientist Paul Frissen (2013) said, how much power do you concede to the secret intelligence services in order to protect the individual right to secrecy (secrecy itself is hidden and therefore not transparent). Life in a completely transparent society is equal to the most horrible life one can imagine. Paradoxes do produce interesting insights sometimes. Many philosophers and scientists are fond of them.

Some paradoxes have haunted philosophers of different eras. The paradox of the liar who says, 'I am lying' is such a paradox. Is he telling the truth? Most of us have tried to solve this paradox, but it seems impossible to decide whether the liar is telling the truth or not. A well as being one of the most important opponents of the American involvement in the Vietnam War, British philosopher Bertland Russell (1872–1970) is one of the founders of analytic philosophy. One of his main fields of interest was linguistics. Russell thought he had an answer in his theory of logical types. He distinguished different linguistic levels, as laid out below.

At the bottom level there is reality, the phenomena themselves. Then, on the next level one can say something about such a phenomenon, for example 'It rains.' On the next level one might say that one is lying when one says that it rains, when everybody can see that the sun is shining. Interesting for our paradox, then, is the statement on the subsequent level that one is a liar when one says that one is lying when having said that it rains while the sun is shining. What one is saying now, is that the discussion on being a liar takes place on another level than the one on the statement 'I am lying.' The fact that one might be a liar or not when one says 'I am lying' is on a different linguistic level.

The answer, therefore, to the paradox of the liar would be that one may distinguish two language levels in this paradox and then the paradox is solved. This logical solution has been used in various manners – for example, in order to explain 'double-bind' situations in communications in which an individual (or group) receives two or more conflicting messages, with one negating the other. The double bind occurs when the person (group) cannot solve the inherent dilemma. An example: a mother telling her child: 'You must love me.' Love is spontaneous and the child can only love his mother of his own agreement. This can be made explicit by distinguishing two levels. On the first level, the child believes his mother's order to love her. On the second level, the mother has the power to manipulate the child with this order. In this way, Gregory Bateson makes use of Russell's logical types in his 'Ecology of the mind' and his analysis of Russell's linguistic solution to paradoxes. He began with the theory of logical types and applied it to his own discipline. And this is exactly what I would like to do with the paradoxical situation in our international classroom (Portegies *et al.*, 2014) of tourism studies having English as a logocentric lingua franca, whereas at the same time we need to use the academic richness of all the other national, cultural language-games as well.

The Paradoxical Challenge of International Tourism Studies with English as Lingua Franca

Let us first imagine a hypothetical situation without English as the lingua franca. For the natural sciences this would not be such a big issue: a mathematical problem seems to be the same in Chiang Mai as it is in Stockport. In the social sciences, however, as in the tourism academy, this looks different. If our main study object, for example, is international tourism, then we must realize that different languages have different language traditions, with a certain degree of solipsism in each tradition. Each tradition will have its own icons and standard works that reflect on tourism. Tourism itself means different things in different language-games all over the world. And this is true for many concepts and theories in the humanities. When Foucault published his first works, a typical French discussion of adherents versus opponents filled the public discourse, as often occurs in philosophical France with new thoughts. The French journal *Magazine Littérairé* (1977: 127–128) provides a good impression of such a vehement discussion about a new philosophy, which often ends up with two different camps, pro and contra the philosophy. In this example, Foucault was the main philosopher who created this debate in French public life. Memory of this type of discussion has been incorporated into French tradition. From the moment Foucault is translated into English, much of the richness of this evaluative discussion gets lost in the English translation. The opposite, of course, also is true. The work of Sir Karl Raimund Popper became known in France only after his oeuvre had been translated into French, with the same effects.

Many more examples could be mentioned in order to demonstrate this dilemma. And this situation becomes the more intriguing, the more we introduce into this debate examples from language traditions from all over the world. If you realize the richness of the Chinese tradition, with all their icons, philosophical traditions, and so on, and the uneasiness of their situation in the international academy, the complexity of the situation becomes more demanding. Discussions about the future of international tourism must surely be enormously diverse if you imagine these cultural differences. A mathematical problem is the same in Japan as it is in Canada, but academic discussions on tourism must be as different as tourism itself in these countries. Challenging questions emerge from this awareness like: What does authenticity mean for a Buddhist from Japan? What type of leisure can be researched in China according to Chinese scholars? What does post-communist tourism look like according to Polish academics? And many more.

It is time, now, to introduce English as the international language par excellence, later to be called the lingua franca. But first, I would like to make a comparison utilizing novelists from all over the world who start to write or to be translated in the same American, international language in order to acquire international reputation. As Tim Parks (Clarke, 2016), English writer and translator, states: these writers present a caricature, a stereotypical image, of their own countries in order to attract global attention. An American writer does not have to do much to attract such attention, says Parks, but a Serbian writer has to adapt the content and style of their book if they want to reach the same international audience. This situation has already existed for quite some time and seems inevitably to grow in importance. On the one hand there is a romantic and sometimes nationalist tendency to glorify the richness of the one's tradition and on the other hand there is a growing group of international writers who appeal to the universal themes of a global audience. The *problematic* way out of this paradoxical situation seems to be, according to Parks, a process of internationalization in which stereotypes are being reinforced all the time and the richness of the novelists' own cultural traditions gets lost in the translation for an international audience. Focusing on your own cultural tradition would imply that you would stay close to the richness of your own cultural sources but that you would accept the high degree of solipsism, of a fixation on one's own internal state, that goes with it. But as George Steiner (1975) claimed, to learn a new language is to open a new window, a new perspective, on the world. These new windows seem crucial to making international tourism studies worthwhile. However, then Parks states that, if you try to go beyond your own cultural limits, the richness of your own cultural tradition, you get lost in the translation for an international, American-speaking audience. That is the paradox of Parks. At the end of his article, Parks appeals to a process of self-reflection in his colleague-writers in order to become independent of any obsession with achieving international fame.

Let us accept English as our contemporary lingua franca – after all, it is a blessing to have a common international language – but without forgetting how this came into being within a postcolonial academic context where many voices have been excluded. Let us remain self-reflective, as Parks has advised his international fellow-writers to be, about how this English spoken force-field of international tourism studies came into being. Therefore, it seems necessary to look a little closer at the historical context that constituted the academic field of tourism today. What is the societal background of this force-field?

Academic Developments in a Network-society

The tourism academy develops within the context of a globalizing world. We all have become aware of the complexities of our new global village, our network-society. New social structures stem from a segmentation of the global economy, an international division of labor, informational-based production and consumption, and an increasing diversification worldwide but also within each region. There are several centers and several peripheries. Global and powerful, capitalistic networks interfere with the networks of regions, states and with the international networks already in existence. At the other end of the scale, local networks all over the world still have their persistent influence on everyday life. Much more could be said about the various networks that are interacting between the global and the local. A relevant question, in this context, of course is how people from within these various interacting networks translate all these influences in their perspectives on the world. There are processes of cultural unification, of 'disembedding mechanisms' as Giddens would say (1991), together with revivals of local cultural elements, there are different forms of cultural globalization (Americanization being one of them), and there are varying transnational networks and third cultures. The whole picture is composed of mixtures of pre-modern, modern, post-modern, post-colonial, globalized, de-territorialized and not to be forgotten virtual networks, which influence the different voices, originating from everyday life in these interacting networks.

It is important to recognize this complex network-society for our insight into the international tourism academy. It reminds us of how social sciences came into existence during the 19th century. Nisbett (1968) described a long time ago how sociology emerged within a context of the political and industrial revolution of the nineteenth century and the concomitant tensions between the old, traditional, agrarian society and the new, modern and urban society. Sociology as a new form of knowledge grew out of this tension in order to better understand the 'new problems' in the turmoil of this new society, where a new and modern 'Gesellschaft' (society) opposed the traditional 'Gemeinschaft' (community). One might see sociology as an attempt to understand this new society, whereas cultural anthropology represented the nostalgic attempt to preserve disappearing pre-modern, traditional, rural societies and searched to gratify this nostalgic need in a colonial era by idealizing faraway, agrarian, non-Western cultures, untouched by modern civilization. Sometimes this 'pastoral tendency' (Clifford & Marcus, 1986) still persists in anthropological or tourism discourses.

The point, here, is that – like in the 19th century – today one might also speak of a new, in this case network-society that underlies the need for new knowledge of, and insights into, the new issues that dominate our international tourism scene. How, therefore, does the international tourism academy look as a reaction to this new network-society?

Academic Insulae and the Route to Ithaca

In 'Academic insula: The search for a paradigm in and for tourism studies' (Platenkamp, 2015: 5) I tried to apply the concept of a paradigm to the situation in the tourism academy. Social sciences and the tourism academy never had a paradigm, in Kuhn's understanding of it. For example, maybe in psychology you might call Darwin's theory a paradigm but then you are reducing psychology to biology, a natural science. Therefore, it seems better to present the social sciences and the tourism academy as pre-paradigmatic. Tribe (1997) refers to this situation as the 'In-discipline of Tourism Studies'.

In such a pre-paradigmatic context, as in the situation of national cultural traditions, the different so-called 'paradigms' have some unique qualities that make them incomparable to others. Incommensurability is the word that philosophers use to indicate the logical impossibility of comparing two theories or paradigms. Therefore, when one talks about 'paradigms' one is confronted with the same type of paradox. How should one translate a local book or article into the lingua franca without losing its particular cultural richness? Or how should one organize a dialogue between paradigms, with the same intention?

In my article, I created different islands or pre-paradigms that Odysseus, who stood for the tourism academic, would visit in his journey to Ithaca. There is an island of critical rationalism (positivism, for its opponents), one of interpretivism, an island of critical theory and, then, different islands of national tourism discourse, which are of course of particular interest to our debate. These islands are isolated from each other and they seem to need different 'turns' (like an interpretive or a critical turn) in order to make the academics realize that there might be other islands as well. The question of how to organize a rational discussion between these differing theoretical and national traditions that hide themselves on isolated islands becomes more relevant than ever before in the social sciences and in the tourism academy if one becomes aware of the situation in our network-society. Ithaca stands for the goal of the escape route: out of the paradoxes of these various islands into the international classroom of tourism studies. How does one imagine this island as the home island of Odysseus, where his beautiful Penelope is waiting for him (not exactly a feminist position I admit) and the tourism academy would find a Russellian meta-level on which our terrible paradoxes could be dealt with?

Let us, therefore, situate the complexities of a network-society on Ithaca, first.

Kumasi on Ithaca: The Way Out of the Paradox

Odysseus' travels are symbolical in one's quest for knowledge. In spite of the strong, irrational forces, symbolized by the struggle between the Greek gods, during his journey, this clever hero used his rational capacities in order to mislead the overwhelming influences of the gods. The way in which this can be done in the case of tourism studies can be explored by implementing the basic principles of reading a text as Said (2004) proposed: reception and resistance. Being open to each theoretical tradition while at the same time not losing one's resistance, based on one's own insights and on distrust towards the absolutist, *divine* truth of any insight, is a first way of opening up this discussion.

On each island one can, of course, make an – always incomplete – attempt to receive some important insights through empathy with its background assumptions. At the same time, a critical resistance remains a focal point of interest, that should bring us a step further in the 'intersubjective' dialog between the different 'paradigms' (Guba, 1990). The islands are part of the archipelagos of tourism studies that promise to create the best possible world, that in the end would appear to be our existent (actual) world. However, in this new world it seems necessary to situate the capital of Asante (Ghana), i.e. Kumasi, on Ithaca. Kumasi is our new reality of Ithaca, it symbolizes our contemporary, existent and less and less Western world. It stands for the multi-layered reality of a network-society to which academia has to adapt. Odysseus' dog could guide our hero to Kumasi, this dynamic city, where Appiah (2007) sees an illustration of a contemporary 'cosmopolitan contamination':

> English, German, Chinese, Syrian, Lebanese, Ivorian, Nigerian, Indian: I can find you families of each description (in Kumasi). I can find you Asante people, whose ancestors have lived in this town for centuries, but also Hausa households that have been around for centuries, too. There are people there from all the regions, speaking all the scores of languages of Ghana as well. (Appiah, 2007: 101–102)

When you understand cosmopolitanism well, says Appiah, you think human variety matters 'because people are entitled to the options they need to shape their lives in partnership with others' (Appiah, 2007: 104)

The writing of a very cosmopolitan writer from (African) Latin antiquity, Publius Terentius Afro, quoted by Appiah, was known to Roman *literati* as contaminated. 'Quot homines, tot sententiae', was his observation: 'So many men, so many opinions'. Recently, Salman Rushdie is another eloquent exponent of the same attitude, but now in a network-society, where he said that the novel that occasioned his fatwa:

> celebrates hybridity, impurity, intermingling, the transformation that comes of new and unexpected combinations of human beings, cultures, ideas, politics, movies, songs. It rejoices in mongrelization and fears the absolutism of the Pure. Mélange, hotchpotch, a bit of this and a bit of that is how newness enters the world. (Rushdie quoted in Appiah, 2007: 112)

The description by Amoamo (2011) of the hybridized Maori in New Zealand confirms this image in the tourism academy. Cosmopolitanism was invented by contaminators in various manners, not only as a consequence of modernization. This world of contamination has become our home. Or, as Appiah (2007: 113) claims: 'we do not need, have never needed, in a settled community a homogeneous system of values, in order to have a home'. In a similar way a 'paradigm' in tourism studies suggests an often unconscious warm home for like-minded researchers. These academics stay on their island forever and resist any potential of another place. They will never choose the sea and its adventures, because it contains 'adventure, heroism and the quest for the unknown'. Some tourism scholars dare to leave their island in order to discover the social sciences, not in its absolute pureness but in the richness of its variety. When looking at the subject that a scholar like Hollinshead has treated – worldmaking, soft science, advanced qualitative representations, underrepresentation, narrative approaches, decolonization, fantasmatics just to mention a few of them – this should challenge us to accompany him 'outside our comfort zone'. When a student in tourism management dares to look beyond the 'quantitative horizon' of the positivist academy, she will discover the pragmatic reality of (participant) observations, in-depth interviews, stories and visual methodologies that enrich her horizon in a substantial manner. Managers in the tourism industry will experience the added value of these studies directly and recognize academics as potential allies instead of villagers of a faraway island, still untouched by cosmopolitan contamination. In general, looking at a network-society in which pre-modern, modern and post-modern elements are interfering in various networks, 'paradigms' have come into existence with varying perspectives on these interfering networks as well. According to the principle of cosmopolitan contamination, researchers are not expected to ignore the place of their origins or their 'paradigms', but, at the same time, they fear the absolutism of the Pure, which is part of this paradigmatic situation, and they are eager to leave their homes in search for new insights that might be enriching.

Knowledge Production in Kumasi on Ithaca

But how do we organize these inter-perspectival dialogs between 'paradigms'? How do we escape from the paradoxes? Cosmopolitan contamination looks like a meta-level that Russell might accept as an outcome. If the ultimate 'paradigm' on Ithaca is represented by the network-society of Kumasi, can we say something about the basic rules in this network-study of tourism? The most important tourism magazines, to start with, would be smart if they would move their headquarters to places like Kumasi. But more important: what does cosmopolitan contamination look like in tourism studies on Ithaca?

Looking at the contamination part of this question, it seems necessary to give the floor to the humanities and their strong attention to interpretive power in grasping the needed context – as the source of this contamination – in a research project. Much more attention to contextualization is needed. For example, in the field of cultural heritage many stakeholders often have their own perspectives that color their expectations. Levuka, the former capital of the Fiji Islands (Fisher, 2004) illustrates this case. Different groups need to be understood in order to manage the cultural heritage of this old colonial city. More generally speaking, many interpretive understandings in our hybrid network-society are in growing need of mutual interpretations. This process of interpretation is captured in a 'hermeneutical circle' (Gadamer, 1990) in which all participants constantly interpret each other's interpretations. Subjective biases are taken, here, as a rich source of information that should be made explicit from within the 'Fragehorizont' (Gadamer, 1990) or 'question horizon' of the participants in this process. In Levuka, there are three groups with their own interpretations of the heritage of Levuka which should be involved in the construction of meaning of this former colonial town:

(1) The ethnic Fijians, for whom the colonial buildings in Levuka have no particular meaning. Places have extrinsic and intrinsic meaning (mana), not buildings.
(2) The 'old' Europeans from colonial times, who see these buildings as landmarks.
(3) The Indo-Fijian and Chinese shopkeepers, who have some main shops in the city and whose prime interest is to earn money. They are afraid of the tourism industry that will profit from the heritage.

All three are to be included in a more refined (hermeneutical) understanding of this cultural heritage. At the same time, the cosmopolitan part of our question tries to identify universals that might be useful for the common concept of cultural heritage. This position of 'distanciation' (Ricoeur, 1981) is not the privileged position of the objectivist scientist alone, but also belongs to the reflections of all parties involved, who try to identify universals in relation to this situation of cultural heritage. There are more 'contextualized' or 'cosmopolitan contaminated' observers in the context of tourism studies. Voices often not heard from, need to be included and a polyphonic dialog between them could lead to higher quality insights and solutions.

The fact that many tourism researchers have doubts about the academy because of the in-discipline of tourism studies (Tribe, 1997) might also be seen as an advantage for the field. In tourism studies, as in tourism management, we may look beyond the limits of the so-called 'paradigms' of social scientific disciplines. Tourism studies already has a tradition of interdisciplinary research from its beginnings. This tradition also enables a more flexible response to the multi-layered reality of the academic and managerial situation. A limitation to different 'paradigms' or islands would imply a brutal stop in the unending and promising quest that tourism studies should be. Tourism studies as a multi-layered field of research offers the opportunity for a flexible, but persistent, discussion between different so-called 'paradigms' from various sources. Let us not bother about the paradigmatic status of tourism studies and let us profit from the freedom that this rejection of the paradigmatic status will create in tourism research. A sophisticated direction of loosely structuring, this new situation in tourism studies could be the distinction between three modes of knowledge production. In social sciences, as in tourism studies, all three modes, as will be explained below, are characterized by an organized form of criticism, pluralism, and reception and resistance. Probably, there will never be a paradigm on Ithaca in tourism studies but, as a 'regulative principle' – the meta-level Bertrand Russell asks for – Ithaca promises a discussion in the whole archipelago that will disturb all the 'paradigmatic believers' who stay on their different islands.

Three Modes of Knowledge Production on Ithaca

The distinction between three modes of knowledge production can be another tool to realize this meta-level in order to escape paradoxes. The distinction could provide us with the structure of argumentation and the criteria that go with it on this meta-level. This structure of argumentation has changed over the last centuries. From a historical (Western) perspective, globally three phases can be distinguished in the production of knowledge. First of all, these phases are based on the relations between academic, professional and normative or more philosophical knowledge. These relations have changed in character from a predominance of religion, via a predominance of science to a more egalitarian relation between the three areas in contemporary network-society.

In this hybridized and global network-society, since the end of the 20th century and alongside the 'traditional' mode of knowledge production (mode 1 knowledge) a mode 2 knowledge has emerged, created in a broader, transdisciplinary social and economic context of application. It has been created because conventional terms such as applied science, technological research, or research development were judged inadequate. It is a new production of knowledge that has a strong influence on the dominating presence of scientific knowledge that has always been interpreted as mode 1. The revolutionary contribution to knowledge development in the applied context of information communication & technology (ICT) during the nineties is a good example. Mode 2 takes place predominantly outside university structures. Criteria of good knowledge are related to sophisticated solutions of complex practical problems, which is different from the truth-finding criteria of mode 1 (see Isaac & Platenkamp, 2012).

As indicated in Figure 8.1 the three modes are to be explained as follows. In mode 1 of tourism studies there will be an unending discussion about the contributions of the natural sciences, the social sciences and the humanities. The main principle, here, is inclusion and not exclusion, although the discussion on academic criteria will never end. A crucial question for this inclusion is this:

> How, in pluralistic societies with a diverse ethnic mix (in a creolizing world, VP) is it possible to narrate histories that include all constituent variants equitably? (Dann & Seaton, 2001: 25)

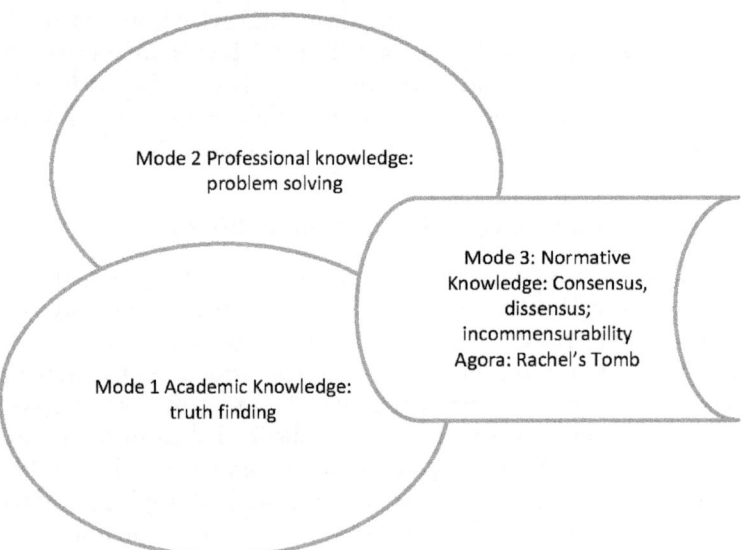

Figure 8.1 Three modes of knowledge production

This question refers to a meta-level that enables one to escape from the prison of one's own background. A mix of different backgrounds is to be taken into account to go beyond one's own cultural limits in order to improve the explanations and interpretations that we are looking for.

The tension between 'explanation' and 'interpretation' in tourism studies is not a problem to be solved but a rich source of new developments for which inclusion and not exclusion is the most justified approach. Various contextualized voices are to be given all the space they need, but subsequently there also should be a universalizing polyphony of the value of all these voices and their confrontations in an organized, critical manner (Platenkamp, 2007).

In mode 1 of knowledge production in the tourism field this contextualized academic discussion contains a very promising challenge that, however, has hardly been stimulated until now. The main reason for this is the parochial domain that each so-called 'paradigm' creates for itself and the ridiculous unbelief that something reasonable could take place on the other islands. Where are the conferences, except for the International Conference on Paradigms in Tourism Studies, held in Finland in 1996, during which different islanders flock together and challenge each other in a constructive and yet critical and contaminated cosmopolitan attitude (Appiah, 2007), of reception and resistance (Said, 2004)? Or should one listen more carefully to the art of sophisticated journalism, where participants follow this path in a more self-evident manner? If one becomes more modest in our scientific pretensions and is not envious about the reputation of natural scientists anymore, then (mixtures of) intellectual traditions can be elaborated and nuanced in a much more challenging way that tries to give an answer to the complicated questions of a network-society. Kumasi on Ithaca will be one's guideline.

Also, when frustrations threaten a positive outcome, mode 2 of knowledge production requires a more decided contextual approach in order to handle the growing complexity in international research and education in and around tourism development. For example, complex contextual learning processes take place, which are not taken seriously enough within the education and research milieu of the field. Fieldwork components play a crucial part in this education and, in the Tourism Destination Management (TDM) master's program at the Breda University of Applied Sciences in the Netherlands, the fieldwork component 'has become both a pivot and a pillar, feeding the overall tourism curriculum, both in content and design' (Portegies et al., 2009; Portegies et al., 2011; Portegies et al., 2014). This fieldwork is the open space where mode 2 knowledge production takes place:

> ... where students can apply their talents for observation, exploration and making their own discoveries. In that same open space

practitioners share their successes and failures. More importantly, there is room to discuss and exchange uncertainties about current livelihood practices, potential market developments and the unknown. Professionals and entrepreneurs of all sorts and sizes play an important role in the fieldwork. These people practically cooperate with students and lecturers as part of an experience of 'learning on the spot', constantly aware of the uncertainties and of the limits of knowledge. (Portegies *et al.*, 2014: 348)

In this process there is a growing accent on post-disciplinary research, as well as a deepening of the study of de- and re-contextualization practices. This refers to a 'collaborative effort of academics from various disciplines and practitioners from many sides who jointly and without pre-established hierarchy are working in an innovative manner with complex and emergent practices by focusing on both context-dependent and context-independent characteristics' (Portegies *et al.*, 2014: 349). In this pragmatic design for contextual learning, more intense attention was paid to the focused interests of stakeholders in tourism destinations. For example, in Bali 'small business operates within a context of larger interests' (Portegies *et al.*, 2014). In this climate, small, independent food outlets were tolerated 'as long as they didn't sell beer (alcohol) and they conformed with what the highly competitive syndicates dictated' (Portegies *et al.*, 2014: 351). This type of information evolved from the repeated conversations of students with the same players in the destination. This inspiring learning context exemplifies how mode 2 knowledge production might be organized in diverse tourism contexts. Here too, Kumasi on Ithaca offers the inspiring image of a respectful, free and (self-)critical knowledge production in the professional tourism contexts of a network-society.

Mode 3 as an emergent form of knowledge production in tourism could play a crucial role here, too (Isaac, 2010). It has been introduced (Kunneman, 2005; Platenkamp, 2007) in awareness that there is a long-term tendency in modes 1 and 2 to exclude the 'slow questions', related to sickness, death, repression but also to moral virtues such as compassion, inner strength, or wisdom and other sources of existential fulfillment that are also included in these slow questions and that remain crucial for all generations in a variety of places. For example:

... in the reflections of tourism developers in Burma, moral questions that are related to injustice, human rights, and the everyday life of local people are excluded in their context of application. As a consequence, original villages have been destroyed, local people removed, and human rights violated for the sake of tourism development. Professionals who abstract from these circumstances do not consider this moral aspect, but focus only on the viability of tourism business. (Isaac & Platenkamp, 2012: 178)

Economic and political power constellations, but also dogmatically defended frames of interpretation, throw obstacles to the necessary development of learning processes in mode 3. Therefore, a relative autonomous development of mode 3 should be claimed that supports more adequate interventions in tourism professional practices. *Dissensus and incommensurability (the logical disconnectedness of different positions) is as possible in the argumentation structure of this mode as consensus is.* This implies that all participants endure differences of opinion, also when they cannot be brought to an agreement. In the hybridized network-society of Kumasi on Ithaca consensus is not the necessary outcome of an ethical discussion between various 'transcendent' values and perspectives. Values need not be but can be incommensurable and in mode 3 these differences are included by understanding them as the expression of a plurality of perspectives. In this discussion there is no Archimedean last point of anchorage but a 'horizontal, transcendent orientation in which there is openness, receptivity and criticism' (Platenkamp, 2007: 45). The orientation is transcendent because it transcends empirical reality and relates to a debate about values. The orientation is horizontal because it does not imply a coercive hierarchy. A clear example of such a discussion has been presented in an article by Isaac et al. (2012) that questions the alleged neutral objectivity in social scientific discussions through the relevant example of how academics concealed their positions of neutrality just before the 86th annual tourism conference of the Organisation for Economic Cooperation and Development (OECD) in Jerusalem in October 2010. The article analyzed 'the relatively high amount of email reactions to a Palestinian tourism scholar who called for support from the tourism academic community for the rejection of Jerusalem as the place where the conference would be held' (Isaac et al., 2012: 159). A mode 3 discussion has been suggested here in order to stimulate a climate 'of broader enlightenment that ultimately goes beyond the perspectives of individual parties' (Isaac et al., 2012: 165). Here, too, reception and resistance remain relevant, but also the critical attempt to replace the argument of power by the power of argumentation, as it suits well in an intellectual tradition where moral discussions do not consist only of expert exposés that analyze the moral traditions in different times and places but also consist of taking well-argued sides in the awareness that the other might always be right. Ethical discussions come closer but also become more contextualized in a hybridized, network-society. In Kumasi on the agora of Ithaca slow questions are taken into serious account.

Conclusion

There seems to be a way out of the difficult paradoxes within the international tourism academy. The lovers of Penelope have disappeared from the island and here we are with Odysseus in his new network-society and its hybridized cultures. English is the lingua franca, but probably we should follow the advice of Ray Boland and Rose de Vrieze-McBean in their presentation on English as the lingua franca, or ELF as they call it (Boland & de Vrieze-McBean, 2014). They refer to an article in the business supplement of the *Observer* newspaper which explained how Korean Airlines decided to buy its flight simulators from a French supplier partly because the English used was easier to understand than that of its UK competitor. Native speakers are more difficult to understand and this goes further than only pronunciation. For example, 'to lose face' has become accepted as Standard English, but 'to give face' has not, although it would be perfectly acceptable in an Asian context as it reflects Asian values. Meaning, let us not forget, results from shared agreement between users. Boland and de Vrieze-McBean, recommend not teaching non-native speakers English or American forms of politeness, etc. it is a waste of time. Trying to teach Chinese people to be like Americans will fail, they say. Successful communication in business English requires practicality and an awareness of context and they provide many examples in their article, which calls for global instead of the Queen's or American English. When students learn ELF they will experience many varieties of English and different communication strategies that go with them. It is an interesting, decentering field in development.

Apart from a lingua franca, one can also speak of a cultura franca as a hybrid cultural academy. This term refers, amongst others, to the latent voices that are often hidden in the background assumptions of participants in contexts that do not come to the surface. When the voices in international tourism destinations come from so many and variegated cultural backgrounds, they need to be understood in a new approach that supports a meta-level of argumentation that lifts us out of our own cultural limitations.

This perspectival clash could be organized in a polyphonic dialogue on the 'agora' of Hannah Arendt. Basic principles in this dialogue, like cosmopolitan contamination, reception and resistance, or the type of argumentation in the three different modes, have already been mentioned. It could lead to the decentering of the English and Western-centered international tourism academy and at the same time make use of the opportunities of English as a lingua franca. World-making becomes a challenging opportunity from within various language traditions that at the same time attempt to communicate in one lingua franca.

Landmarks, intellectual traditions, economic, political and social insights, are to be translated in such a way that the richness of them would be further developed instead of excluded. To organize such a polyphonic dialogue seems to me to be a gigantic opportunity for the future of this academy. Probably, in the end, paradoxes will not be solved definitely: they are too strong and mercurial for that. Does the liar lie or is he speaking the truth? remains a paradoxical problem to be solved in many contexts. Even Russell did not prevent paradoxes from popping up after he thought of his logical types. We might say that life itself is one big paradox. We are thinking and planning as if there is no end to it, as if our lives are infinite while at the same time, of course, we cannot deny the finitude of human life. Philosophy has struggled with this paradox for ages and has formulated many, often inspiring, answers to it.

In the same way, Russell's logical types could inspire us to new and beautiful landscapes in the international tourism academia. I have tried to show you such a landscape – or should I say, using an ugly word, a meta-landscape today.

References

Amoamo, M. (2011) Tourism and hybridity: Revisiting Bhabha's third space. *Annals of Tourism* 38 (4), 1254–1273.
Appiah, K.A. (2007) *Cosmopolitanism: Ethics in a World of Strangers.* London: Penguin.
Aster, E. von (1968) *Geschichte der Philosophie.* Stuttgart: Alfred Kroner Verlag.
Bateson, G. (1972) *Steps to an Ecology of Mind: Collected Essays in Anthropology, Psychiatry, Evolution and Epistemology.* Chicago, IL: University of Chicago Press.
Boland, R. and de Vrieze-McBean, E.R. (2014) English as Lingua Franca. Presentation at the Education Day of Academy for Tourism, 22 February 2014. NHTV Breda University of Applied Sciences.
Clarke, J. (2016) Without illusions: Jonathan Clarke interviews. Interview by Tim Parks. *Los Angeles Review of Books* 6 July.
Clifford, J. and Marcus, G. (1986) *Writing Culture: The Poetics and Politics of Ethnography.* Berkeley, CA: University of California Press.
Dann, G.M.S. and Seaton, A.V. (eds) (2001) *Slavery, Contested Heritage and Thanatourism.* New York, NY: Haworth Hospitality Press.
Fisher, D. (2004) A colonial time for new colonial tourism. In C.M. Hall and H. Tucker (eds) *Tourism and Postcolonialism: Contested Discourses, Identities and Representations* (pp. 126–139). London: Routledge.
Frissen, P. (2013) *De Fatale Staat.* Amsterdam: Van Gennep.
Gadamer, H. (1990) *Hermenutik I: Wahrheit und Methode.* Tubingen: JCB Mohr (Paul Siebeck).
Giddens, A. (1991) *The Consequences of Modernity.* Cambridge: Polity Press.
Guba, E.G. (ed.) (1990) *The Paradigm Dialogue.* London: Sage.
Isaac, R.K. (2010) Moving from pilgrimage to responsible tourism: The case of Palestine. *Current Issues in Tourism* 13 (6), 579–590.
Isaac, R. and Platenkamp V. (2012) Ethnography of hope in extreme places: Arendt's agora in controversial, tourism destinations. *Tourism, Culture and Communication* 12 (2/3), 173–187.
Isaac, R.K., Platenkamp, V. and Çakmak, E. (2012) Message from paradise: Critical reflections on the tourism academy in Jerusalem. *Tourism, Culture & Communication* 12 (2–3), 159–173.
Kunneman, H. (2005) *Voorbij de dikke-ik.* Amsterdam: B.V. Uigevrij SVP.
Magazine Littéraire (1977) *Foucault.* September, vols 127/128. Paris: Du Scorpion.
Nisbett, R. (1968) *The Sociological Tradition.* Berkeley, CA: University of California Press.
Platenkamp, V. (2007) *Contexts in Tourism and Leisure Studies: A Cross-cultural Contribution to the Production of Knowledge.* Wageningen, Netherlands: Wageningen University Press.
Platenkamp, V. (2015) Academic insula: The search for a paradigm in and for tourism studies. *Tourism Analysis* 5, 561–571.

Popper, K. (1945) *The Open Society and its Enemies. Volume 1*. London: Routledge & Kegan Paul.
Portegies, A., de Haan, T. and Platenkamp, V. (2009) Knowledge production in tourism: The evaluation of contextual learning in destination studies. *Tourism Analysis* 14 (4), 523–536.
Portegies, A., de Haan, T., Isaac, R. and Roovers, L. (2011) Undertsanding Cambodian tourism development. *Tourism, Culture and Communication* 11 (2), 103–116.
Portegies, A., Platenkamp, V. and de Haan, T. (2014) Embedded research: A pragmatic design for contextual learning – from fieldtrip to fieldwork to field research in Australasia. In D. Dredge, D. Airey and M. Gross (eds) *The Routledge Handbook of Tourism and Hospitality Education* (pp. 348–355). London & New York: Routledge.
Ricoeur, P. (1981) *Hermeneutics and the Human Sciences*. Cambridge: Cambridge University Press.
Russell, B. (1903) Appendix B: The doctrine of types. In B. Russell (ed.) *The Principle of Mathematics* (pp. 523–528). Cambridge: Cambridge University Press.
Said, E. (2004) *Humanism and Democratic Criticism*. New York: Palgrave/Macmillan.
Steiner, G. (1975) *After Babel: Aspects of Language and Translation*. Oxford: Oxford University Press.
Tribe, J. (1997) The indiscipline of tourism. *Annals of Tourism Research* 24 (3), 638–657.

9 The Call for 'Dynamic Genesis' (after Deleuze) in Tourism Studies

Keith Hollinshead, Rukeya Suleman,
Sisi Wang, Bipithalal Balakrishnan Nair
and Alfred Bigboy Vellah

Introduction: Tourism and the Hidden and Unexpected Truths of our Time

We have learnt from many authors such as Buck (1993), Rothman (1998), Thomas (1994), McKay (2004) that tourism is a very important industry because it inscribes what peoples, places, pasts and presents are (or are supposed to be). We have learnt from Kirschenblatt-Gimblett (1998) that there is an epistemological logic to tourism, which potentially influences how places and spaces are 'made' (i.e. the *madeness* of destinations, in her terms) and rendered distinct as being of a particular 'here' (i.e. the *herenesses* of localities, in her syntax). We have learnt from Horne (1992) that tourism is harnessed by groups/communities/nations to sanctify, to celebrate, or to triumphalise vision 'x' over culture and/or nature in contestation with vision 'y' (over similar). We have also learnt from Hollinshead and Suleman (2018) that all across the world individuals and institutions deploy tourism in worldmaking fashion to legitimate preferred versions of being/becoming vis-à-vis alternative versions. Such is the power of tourism to frame, to structure, and to conform.

The aim of this chapter is to explore this worldmaking, or projective, agency of tourism to mainstream particular vistas of identity or aspiration for particular peoples. We also consider how tourism sites/sights/storylines/scenarios are harnessed to articulate preferred interpretations of 'culture', 'history', and 'nature'. To this end, our aim is to explore how those populations can/could performatively manage those vistas as vehicles of preferential narratives of culture, history, nature, etc. (see, here, Hollinshead's (1999) critique of Hornes's concept of intelligent tourism, and of the accordant commonplace *ideological legerdemain of tourism*). In probing these powerful representational acts, the purpose of this chapter is to draw attention to the often hidden and unexpected – or rather the unexamined/underexamined – power of tourism to pre-interpret the world for us whether we are (in the given instance), local citizens of destination 'a', or intrepid travellers in search of storyline 'b', or of experience 'c', perhaps.

In carrying out this critical exploration of the power of tourism to authorise particular narratives of culture, history, and the nature of places, the chapter will examine some of the conceivable Deleuzian paradoxes of and about tourism (that is, of Deleuzian vistas translated to the industry of tourism and the field of Tourism Studies) which might run counter to received opinion, or seemingly antithetical to routine understanding, or otherwise opposed to certain forms of held commonsense. To this end, this quest for paradoxes – where the term 'paradox' derives from the Greek word *paradoxon*, meaning contrary to expectation (Soanes & Stevenson, 2006: 1275) – constitutes a quest for that which is seemingly beyond belief or 'unbelievable'. In order to synthesise such found paradoxes, the French philosophiser Deleuze will be harnessed to help critique not only who is apparently doing what to whom and which through tourism, but (more significantly) how else might peoples, places, pasts or presents otherwise be inscribed/projected/declared through tourism. And thus, Deleuze will be brought on board ontologically whereby ontology (in the paradoxical and non-prescriptive Deleuzian sense) comprises not what *is already created* or *what first appears to be true*, but what *may be* or what *could be discovered* (May, 2008: 15–17). Such will be the conceptual service of Deleuze, 'the arch crusader against commonsense' (Colebrook, 2003: back cover), the lead ethicist of and about not only the dynamisms of 'thinking' but the dead-end-zones of 'thinking' (Williams, 2009: 194–201).

Deleuze and Tourism Studies Refreshed

In many ways, the mercurial work of Deleuze augments the insights of Horne (1992) on the general limitations of understandings that characterise the industry of tourism management and development today, an industrial and service domain that Horne deems regularly to be prosaic, superficial and non-creative. And in like fashion, the eclectic thoughtlines of Deleuze augment the judgement of Robinson and Jamal (2009: 696–698) on the general lack of depth of understandings that characterise the academic field of Tourism Studies, which Robinson and Jamal considered to be unadventurous in its cross-disciplinary and transdisciplinary accounts of who is doing what to whom, where, when, why, and how, through tourism. Hence, a call is made in this chapter to encourage much more frequent use of Deleuzian thought in and across the field. It is thereby hoped that those who work in tourism management/Development – and those who research this and that through Tourism

Studies – can overcome current constraints of thinking that continue to privilege rather doctrinaire views that tourism is fundamentally (and almost only!) a matter of business and economics. Deleuze was ardent in his dislike of the presence of such longstanding dogma within any academic field. Thus, the goal here is to bring on board the Deleuzian distrust of entrenched dogma (in this particular instance) to overturn dictatorial assessments of tourism built upon hackneyed and inflexible cognitions of identity and difference. In this light, perhaps Deleuzian thought can be helpful in moving Tourism Studies further towards enriched conceptual understandings about what meets which at travel 'encounters' and about which affects what at tourism 'events'. And all along the journey, we will learn that 'events' (viz. the irruptive and potentia-generating *event*) is a key concept for Deleuze (see Colebrook, 2006: 102).

Background: Deleuze and the Dogmatic Image of Thought

In order to assess how Deleuzian concepts can be employed in relation to tourism, it is constructive to provide insight into Deleuze's developed insights into 'thinking', per se. Principally, Deleuze is a philosopher who probes how one might live in a richer and more creative fashion. Although he did not expressly write about tourism, ipso facto, one may surmise that he would regard 'tourism'/'travel' akin to the way he regarded art – that is, as a vibratory force that can enliven life once it is recognised not so much as a strong and resolute industry or business (or in favoured terms 'a molar machine' – see Deleuze & Guattari, 1987), but as a pulsating stimulus that does not just represent life but which can open up understandings about and possibilities for it.

For Deleuze, art is a realm of activity, endeavour and experience that can draw people out to unthought understandings, and open them out to 'life'. In this sense, art acts as an outside power that can provoke people to a grasp of the world beyond the known and already-recognised. Art might on one level appear to present 'an actuality' but its mightiest strength is its capacity to show and inspire multiple styles and fresh creations/fresh understandings of and about the world, especially where minoritarian outlooks/minoritarian visions are signposted to disturb comfortably established majoritarian reproductions of the world. In this regard, the effective function of art is not 'representational' (Deleuze & Guattari, 1994: 193), ipso facto, for it does not rely upon pre-perceived subjects or on already classified traditions but augurs virtual possibilities in and of the world instead. The Deleuzian value of art is that it can free people from being contained within finite and made-manageable interpretations of being to unleash/free-up/restore infinite interpretations of becoming so that the artist is he/she who adds 'new varieties'/'new vistas'/'new visibleness' to and about the world (Deleuze & Guattari, 1994: 175–202).

Just as art can help us realise how the world has already been 'made present' or 'disclosed' to us institutionally, it is the view of the authors of this chapter that tourism (via both the naturalising activity of the creative host-declarative-agency and the sensation-gaining experiences of the observing traveller) can inculcate new perceptions/new feelings/new thoughtlines about the world that are rather free of received grounded understandings and foundational subjects. Thus tourism can potentially un-discipline us and function like art in enabling us to see beyond what Deleuze deemed to be 'habitual relations of recognition' by encouraging and enabling 'us' to think over and above established forms of temporalisation and spatialisation (see Colebrook, 2006: 106). Ergo, just as art has the capacity to open up fresh contexts and new worlds, so tourism can help transform and recreate the teeming potentiality of life. If art can *counter-actualise* towards new planes of composition, so tourism can also *counter-actualise* towards new understandings. If art can free-up thought via the provocation of new affects and sensations (Colebrook, 2006: 114), then so can tourism potentially and creatively help us see and think differently about the world-uncircumscribed and towards boundless reaches of becoming. The problem is that Deleuze (with or without Guattari) never travelled conceptually into 'tourism', per se, so preoccupied was he (they) with the affect-productivity and the incited-multiplicities of the arts, music, and the cinema.

What is important to Deleuze – whether he is observing the arts, tourism, or whatever – is the capacity of individuals to release themselves from their own imagination, or rather to free themselves from their received figurations (i.e. from the *en groupe* or institutional prefigurations they sustain) about life. To Deleuze, the width and breadth of life lies beyond actual human perception, and there is gross and perpetual danger in the overcoding of particular *subjects* by overcoding groups or bodies of *humans*. A more useful philosophy of and for life, according to Deleuze, is one which refuses to elevate single and secure subjects or single and secure images above other possibilities of understanding, and (instead) accepts the emergent *immanence* of fresh things or newly-invigorated images as the flow of life constantly provides opportunity for novel encounters and verdant experiences. In this light – and heavily influenced by Spinoza (1632–1677) – Deleuze works with a philosophy of existence that suggests life is best lived when individuals remain perpetually *open* to other experiences, to other relations, and to other connectivities. He is, therefore, a conceptualist of 'becoming' (Colebrook, 2003: 125–145) rather than one of 'being' (i.e. he is one who is always alive to the prospects of metamorphosis rather than one safe and sheltered within an 'already-established-being').

Thus, to Deleuze, it is not hardened *nouns* that count, it is flexible *verbs* (viz. infinitives) that matter, since verbs tend to be axiomatically open to the potential for new experiences, the possibilities of fresh relations, the potency of invigorating connectivities. Thus, Deleuze would implicitly be interested in the aspirational value of tourism - that is, in its pregnant power to affect (or provide liberating experiences which disrupt) the everyday ways or the received styles in which the world has hitherto been known in any given place. To Deleuze, the value of art and cinema lies in the capacity of those activities to disengage habitual flows of experience: one may inherently presume, thereby, that he would see 'tourism'/'travel' in the same hue, viz. as a happening or an event (that vital word again!) which can unsettle a or 'the unified thinking subject' (Colebrook, 2003: 24). And the function of tourism would parallel the function of art and of cinema (and also of music) to extend not just the power of thought but the power of that thought *beyond the human point of view*.

Given these preliminary understandings about Deleuze (and about Deleuze in tandem with his co-conceptualist, Guattari), it soon becomes rather easy to recognise him as a paradoxical thinker extraordinaire. If paradoxes are the stuff of incongruity, or the capacity of a statement/a concept/an idea to take us outside our usual, our commonplace, our accepted way of thinking, then Deleuze is indeed an arch paradoxist. Deleuze is a master of the seemingly incongruous, a conceptualist whose statements at first seem bedevilled within the contradictory or under the absurd, but which (upon due reflection) commonly express interesting-new-truths. Consonantly, Deleuze is a philosopher whose assessments regularly conflict with common beliefs, and whose performances appear opposed to existing notions about life. And so, we are at first astonished (paradoxically) when Deleuze informs us that, for instance:

- it is not one's identity that counts but *the intensity* with which one holds a or any particular identity (Colebrook, 2003: 2–3), since potentially there is a plenitude of 'actual' and 'virtual' realities available to each of us;
- it is unhealthy to only ever invest in 'already-founded' or 'actual' subjects, for *virtual realities are not just add-on realities* but are vital arenas of and for 'becoming' that ought to be sought/recognised/ maximised by each of us (May, 2008: 48);
- it is wise to recognise that aspirationally the future is 'open' and 'free', but if we only ever look towards it from a single/fixed/resolute point of view, we will never grasp what 'it' indeed *is* (Colebrook, 2002: 174– 177), and the future does not come in readily-findable 'steps' but via attendance to (and 'play' with) other emergent lines of flight;
- it is judicious to examine what we do to ourselves through the rampant capitalism of our time; capitalism may indeed be a dangerous phenomenon for human beings as its consumerisms govern our desires and regulate what is seemingly permissible for us these days. But to Deleuze, capitalism (or rather 'financial capital', itself) is no

overarching code or belief (i.e. something which only impedes us) and 'it can indeed be integrated positively/productively within our sought desires and our emergent aspirations' (Colebrook, 2002: 128). Hence, to Deleuze, capitalism has no singular and crushing external authority 'that must be obeyed'!

An inspection of Deleuze's writing thereby informs us that he does not celebrate 'knowledge' so much as 'the unthought' or 'the not-yet-thought'. In his quest for immanence of and about 'things' and their fresh relationships or their different connectivity with other 'things', he sought liberation from fixed images of thought, and warned against the worship of knowledge that only arises from already-constituted forms/already-established entities/already-assured subjects. To Deleuze, humans are inclined to be blind to the originating conditions that created those forms, those entities, those subjects and he perpetually rallied against rigid lines of knowledge that were institutionally pre-implanted via *dogmatic images of thought* (May, 2008: 135), be it via tyrannising structures of Western thought, or be it via whatever other overcoded predisposition. To Deleuze, individuals need release from such arbitrary and arrogant 'assertive knowledges', for what counts (to him) is not the conceptual stability of 'a comprehensive knowing' about anything, but how encountered things always continue to palpate against each other and thereby (potentially) everlastingly create new insights/fresh understandings/different possibilities for life (May, 2008: 20). What matters, to Deleuze, is not the correct image of a particular something, but the invigorating/unseasoned/unpolluted possibilities that vivid palpation between different entities/ideas/lines of flight might indeed bring about. What is important (to him) is not the true image of the known (i.e. of pre-known) subject, but the raw seeing-sensation that this *palpation* – i.e. this continual happenstance-pulsation and this admixed transformative-vibration – indeed accords (Colebrook, 2006: 102).

In Deleuze's hands, philosophy does not seek to offer a coherent framework from within which we can see ourselves and our world whole. It does *not* put everything in its place. It does not tell us who we are or what we ought to do. [To him,] philosophy does not settle things. It disturbs them. [To Deleuze,] philosophy disturbs by moving beneath the stable world of identities to a world of difference that at once produces those identities and shows them to be little more than 'the froth of what there is' (May, 2008: 19). Deleuze's philosophy is therefore asystemic (Deleuze, 1994) in that he looks for responses to affect, rather than pursuing meaning, per se, and where he calls for fixed and implanted thought to be 'freed up' from over-defined or overdetermined terms that have lived on well beyond their situational usefulness.

Thus, seeking liberation in and of 'thought', and not adamantine 'knowledge', per se, Deleuze's complaint is that people today (within their institutions/organisations/fields/disciplines) have stopped thinking and experiencing, and must vitally be encouraged towards forms of 'disturbed' or freshly 'disarranged' or 'creative' thinking, or otherwise thinking is not 'productive thinking'. To him, the unfixed relationality between things matters considerably, and the value of activities like art/cinema/music is that they are loaded with the potential for 'affect' (i.e. to reach and affect, or to be impinged upon and be affected, or otherwise to co-influence and affect other 'things') and thereby possess the budding wherewithal to reconfigure the world (Colebrook, 2006: 105). If people are to decently appreciate the wider world and the capacious cosmos of which they are part, dogmatic images of thought built upon human predilections are thereby highly dangerous understandings. Hence, the throbbing multiplicity of surrounding durations here, there, and out-there ought to be more generally (and productively cum profitably) recognised.

In this regard – if Deleuze had indeed ruminated upon tourism and travel – he would no doubt have identified the power of the fortuitously incursive 'event' that is intrinsically bound up in and of each tourism projection and each travel encounter. Just as art/cinema/music can themselves affect, disturb, and freshly palpate understandings, so tourism can constitute an important and loaded latent vehicle of productive disturbance, which (to Deleuze) is *necessary disturbance*.

Deleuze's notion of the rhizomatics of events and encounters adds further useful illustration. For him, events and encounters, by definition, are random, decentred, and proliferating (Colebrook, 2002: xxvii). Whilst, in their classification of and about the world, researchers often tend inherently to look for finite/enduring structures and for coherent/predictable relationships between things (i.e. prescribed subjects and situated objects), in contrast, rhizomatics is the Deleuzian recognition that some 'things' relate to each other on much looser bases. It is the rhizomatic that can operate without hierarchy from any position or point of connection as the force of phenomena mesh and transform with other influences by way of shallow resonances that can break at any juncture and that can fast-connect/readily-reconnect 'elsewhere' (Bogue, 1990: 125). Hence for Deleuze (and for Deleuze and Guattari under Deleuzoguattarian cartography (see Bonta & Prontevi, 2012: 25)), it is the unfilled potential of things that excites as new futures beckon for the rhizomatic 'happening'. In a sense, it is *the disruptive* that can transform old knowledge and beget new potentialities: paradoxically, it is the disruptive and the interfering that generates 'problems' and it is these problems that yield generative movements and productive responses.

In relation to tourism, for example, we might, or perhaps should, carefully distinguish between old ideas about the hackneyed *impacts of tourism* (vis-à-vis economic impacts, social impacts, environmental impacts, etc.), which early theories in Tourism Studies assumed typified what was going on at each tourist destination or place/space caught up in concentrated tourism development, and *Deleuzian rhizomes*, which are emergent or unfolding activities/amalgams that arrive unpredictably (from where?) at a particular point/event/intersection and seemingly travel on unpredictably (to where?). In early Tourism Studies literature, the power of the so-called *impacts* was assumed to be largely generated externally and unidirectionally, and the influence of 'internal agencies' (at the host destination/place) or of bidirectional/diagonal/multidirectional 'activity' (about and around that destination/place, rather than of impacts, per se) tended to be heavily under-gauged (see Lanfant, 1995). The Deleuzian concept of rhizomatics takes these matters of diagonality and multidirectionality to whole new levels of event-hood and engagement, across a whirlpool of many possible dimensions. Within tourism studies, already, Bruner (2005) has reminded us that 'the local' is never a constant nor a steady place with regard to the vicissitudes of globalisation; and Salazar (2010), too, has voiced the view that culture is always something that circulates across populations (where subjects and objects dynamically merge/demerge/re-emerge) in what he terms 'the ever-thickening webs of connectivity' of our time (Salazar, 2010: viii). To Deleuze, life is 'a changing multiplicity of becomings' (Colebrook, 2006: 48), and so Deleuzian questioning might focus, for example, on how the 'unsettling' in any tourism place via the new strong presence of tourism 'there' gives rise to new 'creative' tourist-orientated activities, new storylines, and not-previously-encountered enterprises being animated 'there' and 'elsewhere'.

Deleuze and the 'Opening up' of Tourism Studies

The implication provided by the above background, then, is that a sincere and rigorous engagement with Deleuzian thought can help Tourism Studies become further disinfected here and there from the stranglehold that its predominant orientation to (for instance) business considerations and economic performance have over it. While there have been few treatments of tourism that precisely and openly work to explicit Deleuzian cognitions, a number of commentators in Tourism Studies have implicitly addressed such matters of reimagined life through travel, and immersion in new visions of life through the experiencing of the flow of 'culture'/'nature'/'events'. For instance:

- Horne's (1992) insights in *The Intelligent Tourist* are a substantial call for those who work in tourism management and development to devote more time to understanding how different tourists 'sight-experience' a place rather than just 'sightsee' it, and he calls for richer appreciation of the 'darshana' (Hindi term) for how such vistas are felt and absorbed in the hope that such sights/sites can generate productive forms of 'pradakshina' (another Hindi term) whereby the visitor not only meditates about the spiritual or otherwise special qualities of

the place they are seeing but can think it germinally into evaluations and re-evaluations of their own ongoing general life. Horne's ideas are decidedly Deleuzian in their call for creative becoming.
- Kirshenblatt-Gimblett's (1998) *Destination Culture* examines how exhibited places in tourism can be virtually enhanced via the saturating effects of olfactory, gustatory, auditory, tactile, kinaesthetic, and visual effects, not only in terms of narrative interpretation of the created 'madeness' and 'hereness' (her own two aphoristic terms!!) of those locales and claimed inheritances. And her work on the ethnographic marking of supposed realities and of repetitive placement of virtuality touches regularly upon what Deleuze would deem to be the regularised overdeterminations of industrial tourism.
- Rothman's (1998) lengthy examination of the *Devil's Bargains* of tourism scrutinises the degree to which tourism is a mighty postmodern force that has (notably in tourism-industry-dominant places like Las Vegas) lost meaningful relation to the actual lived world. Rothman is perturbed yet also in thrall to these postmodern presentations and performances of place and space as they are driven by simulacra, or copies and images of things that may or may not have any 'real' association with the locations and destinations which they are deemed to announce and celebrate. In 1998, the Rothman view of the loss of relation to actuality was distinctly Baudrillardian, but could such scrutiny of the 'loss of the real' have been more richly conducted (instead) via Deleuzian insights into the virtual power of becoming for both tourists and their dealing-in-the-unreal hosts?
- Bauman's (2003) work in the Hall and du Gay book, *Questions of Identity*, is a well-noted inspection of tourism as various forms of 'strangerhood'. It examines the journeying of particular types of tourist as they absorb the flux and opportunities of contemporary life to differentially experience the world, and Bauman indeed addresses the flaccidity of the identities of our ever-mobile era. In so many ways, Bauman nuances the acts of belonging and becoming that the Deleuze constructions of rhizomatic influences and lines of flight trajectories address, although he is more concerned about the practical liquidities that wash around human actors today rather than of the posthuman events and encounters that Deleuze is himself drawn towards.
- Mavrič and Urry's (2009) inspection of the New Mobilities Paradigm in Tourism Studies is an informed distillation of the nomadic metaphysics of contemporary life, and while their treatment of the enhanced 'mobile' ways of thinking draws on the performativities addressed by Augé (1995) and Clifford (1997), it is oddly silent on the nomadic thoughtlines of Deleuze/Deleuze and Guattari, per se. But the Mavrič and Urry emphasis on how the travel experiences of people are occasioned via intersecting networks of mobility and hybridised via negotiation through the fresh co-presence of others could indeed

be richly punctuated and leavened via Deleuzian notions of affect, assemblage, and immanence.
- Gibson's (2018) treatment of cultural tourism in Australia is a useful critique of the role of tourism in the 'nationing', that is, in the strategic and normalising construction, of the cultural identity of a country. As a summary view of the nation's cultural production and consumption via tourism, it is a pungent commentary on the demise of the singular (and the overdetermined?) term 'tourism industry', and it is a rare (again implicit rather than explicit) condemnation of 'the limited thinking and ingrained assumptions' (Gibson, 2018: 116) that have characterised the interface between tourism and the creative 'culture-making' industries over recent decades. His judgement (seated in Australia) that 'tourism industry marketing strategies have lapsed back into clichés [and appear] stuck in a rut, relying on a simplistic version of, rather than the remaking of, Australian culture' (Gibson, 2018: 125) is pregnant with Deleuzian overtones/Deleuzoguattarian possibilities. One hopes that any later version of this chapter on creative nationing can indeed pointedly harness Deleuzian conceptualisations on repressive and transcendental thinking (see Colebrook, 2002: ix) vis-à-vis the discovery and invention of new possibilities of national life (after, for instance, the Nietzschean thinking of Deleuze as conveyed within Deleuze, 1983: 101).
- Saxena (2015) has conducted research along Deleuzian lines by inspecting what amounts to 'the dogmatic images of thought' that particular well-placed stakeholders in local and regional tourism indeed champion in their control over tourism development. Her work (after Deleuze and Guattari) on 'the imagined relational capital' that drives small firms which – in England – are active in tourism in the Yorkshire coastal towns of Scarborough, Bridlington, and Whitby, is a most useful pioneering study of the consciousnesses that do and do not propel tourism development. The field can conceivably benefit from an inspired group of informed Saxenas subsequently probing just where the imagination of corporate and allied institutional networks starts and stops and thereby of what they 'do' (i.e. do and do not 'produce').
- Braidotti's (2011, 2013) work in *Nomadic Theory*, and later in *The Posthuman*, is important to follow for those who seek to investigate what tourism is today, what tourism does today, what tourism makes today, and what tourism means today, because it examines our conventional vis-à-vis our emerging understandings of identity and difference, and the drawcards of travel and tourism are indeed principally constructed upon such notions of inscribed identity and viewable difference (Leite & Graburn, 2009: 39/50). Tourism is thereby a domain of accelerated human experience that can perpetually help cultivate new forms of belonging and aspiration at speed via the ordinary play of what Braidotti (2011: 233) might term *its webs of encounter*. Therefore, those who study tourism must learn to monitor many sorts of tensions, contradictions,

and contrarieties in those psychic and symbolic games of becoming. Such are the dynamic and multiple ecologies of belonging that are activated via tourism in general and by what Braidotti (2011: 173) deems to be 'the difference engine of advanced capitalism' in particular. Braidotti (like Deleuze before her) considers the role of capitalism to be distinctly paradoxical in these transformations of self and becoming. On the one hand, capitalism is judged by her (and by Deleuze) to be a pernicious and negative threat to existing forms of social and cultural stability. Yet on the other hand it is saluted as a mix of advanced technologies that can profitably and positively bring about many fresh opportunities of knowing and belonging for people to give them new kinds of validity and refreshing sorts of perpetuation. Consonantly, for Braidotti, tourism itself constitutes a laboratory for previously untried, innovative forms of becoming. Its inherently wide geography beckons new landscapes of relationality as it rather openly triggers unheralded forms of becoming that are somehow rooted, or at least seeded, in what has gone on before. With its everyday plenitude of environmental, cultural, and artistic-aesthetic experiences, tourism *naturally* has a potential superabundance of 'becomings' coursing through it or via it. Its global cartography and the meteorology of heritage/cultural/identificatory forces it routinely rubs up against can readily heighten aesthetic and poetic awarenesses of and about 'the other' for those who have travelled. Such is the generative potential of tourism to declare not just what different pasts/presents/futures have been, or are, or will be, but also how those visions of the other can be (re)imagined.

Summary: The Call for Dynamic Genesis

This chapter has explored some of the views of the metaphysicist Deleuze on the unmet possibilities of experiencing life, or otherwise what Garoian (2015: 491) has termed Deleuze's outlook on 'the entangled ontology' of perpetual differentiation. It has presented Deleuze as something of a maverick thinker who – either working alone or in harmony with his compatriot, the psychoanalyst Guatarri – has oxygenated thought about the relationships between identity and difference. In helping introduce the conceptual infusions of Deleuze into our awareness, many of the contributions of Deleuze on knowledge-making have been found to be loaded with fresh protean insight on matters of being and becoming, by encouraging individuals in all sorts of fields and contexts to creatively and productively overturn the fixities of over-concretised identities and identifications. Put another way – and applied to the industry of tourism and the realm of Tourism Studies (and, to repeat the point, to provinces of knowledge that Deleuze did *not* himself specifically visit or linger in, despite paradoxically the forms of mobility (*nomadic logic*) which he cultivated with Guatarri) – Deleuze was a conceptualist whose insights can considerably help disimagine and reimagine what tourism can produce/can enable/can empower.

While the pivotal question in and across the philosophy of Deleuze is 'How might one live?', an inspection of Deleuzian thought (and of Deleuzoguattarian imperatives) teaches those who work in tourism that the act of travel is an immense field by and through which all manner of new encounters, fresh events, and liberating experiences can indeed be obtained. While in general, Deleuze 'offers a view of the cosmos as a living thing' (May, 2008: back cover), for those who work in tourism his ideas offer a view of a broad *posthuman* world in which all manner of 'durations' can be seen and experienced naturally, spiritually, and culturally in ways that few tourists (and certainly too few management bodies in the industry) appear to have ever dreamed of. Indeed just as in the arts, where Deleuzian thought has been profitably deployed by many organisations and in many nations in the performance of 'a resistance politics' (Finley, 2018) both within the industry and via critical scholastic inquiry, so tourism may be utilised as a distinctly generative differential space for 'human' and for 'human-connected' *flourishing*. If the arts can serve as a multitudinous theatre of (Deleuzian) possibilities (Finley, 2018: 566), so too can tourism/Tourism Studies, once practitioners of tourism management and development learn to appreciate that the industry currently exists not so much as a business projecting rich reflective insight about the possibilities of enhanced life but, rather, as what Giroux (2013) has described as a 'a disimagination machine'. Thus, in these critical respects, Deleuze may be seen to be a principal sage *within the arts* of 'creating beyond' and *within tourism* of 'experiencing beyond'.

For Groys (2008) the phenomenon of *art power* has long existed as a vital force in the dominant plays of local, national, and global politics. No doubt the 21st-century will increasingly see *tourism power* utilised in like fashion. And this will not just be an exercise in *soft power* statecraft authority and jurisdictional privilege (after Joseph Nye, 2004): it will also be an exercise in the creative mobilisation of generative differential encounters, or generative differential 'events'. Such is the promise and latency of *dynamic (Deleuzian) genesis*. But let us not get too dogmatic about it all! It is undoubtedly virtually happening already in so many spaces and places: we just need to tune more decidedly into these infinitus possibilities of flux and of happenstance interactivity. We just do not know everything that is going on, nor everything that could be going on for us, for each of us!

Deleuze is the philosopher of the unthought possibilities in our everyplace travel and in our everyday cherished existence 'at home' within our inherited societies. His thoughtlines constitute an 'instrument for multiplication' of and about the possibilities of life and are a beacon not so much for 'knowing' (certitudinously) but of 'thinking' (potently) (see

Jackson & Mazzei, 2018: 721). Paradoxically, then, the import of Deleuze is not so much what he does for us, but what (in terms of the dogmas that we have been surrounded by but also constrained by) he helps us *undo* for ourselves vis-à-vis our *en groupe* conceptual inheritances. If the central concept of Deleuze's generalised adisciplinary/antidisciplinary work is *the object multiple* (see Mol, 2003), the central concept of Deleuzian thought when applied to Tourism Studies is 'the place multiple'/'the storyline multiple'/'the celebration multiple'. But paradoxically, again, let us not get too carried away here: let us not overcode or overdetermine Deleuze. 'The [Deleuzian] aim is not to rediscover the eternal or the universal, but [for each of us contextually in our own travelled-to and local settings] to find the conditions under which something new is produced' (Deleuze & Parnet, 2007: vii).

References

Augé, M. (1995) *Non-places*. London: Verso.
Bauman, Z. (2003) From pilgrim to tourist – or a short history of identity. In S. Hall and P. du Gay (eds) *Questions of Cultural Identity* (pp. 18–36). London: Sage.
Bogue, R. (1990) *Deleuze and Guattari*. London: Routledge.
Bonta, M. and Protevi, J. (2012) *Deleuze and Geophilosophy: A Guide and Glossary*. Edinburgh: Edinburgh University Press.
Braidotti, R. (2011) *Nomadic Theory: The Portable Rosi Braidotti*. New York, NY: Columbia University Press.
Braidotti, R. (2013) *The Posthuman*. Cambridge: Polity.
Bruner, E. (2005) *Culture on Tour: Ethnographies of Travel*. Chicago, IL: Chicago University Press.
Buck, E. (1993) *Paradise Remade: The Politics of Culture and History in Hawai'i*. Philadelphia, PA.: Temple University Press.
Clifford, J. (1997) *Routes*. Cambridge, MA: Harvard University Press.
Colebrook, C. (2002) *Understanding Deleuze*. Crows Nest, NSW, Australia: Allen & Unwin.
Colebrook, C. (2003) *Gilles Deleuze* (Routledge Critical Thinkers). London: Routledge.
Colebrook, C. (2006) *Deleuze: A Guide for the Perplexed*. London: Continuum.
Deleuze, G. (1983) *Nietzsche and Philosophy*. Translated by H. Tomlinson. New York, NY: Columbia University Press.
Deleuze, G. (1986) *Cinema 1: The Movement Image*. Translated by H. Tomlinson and B. Habberjam. Minneapolis, MN: University of Minnesota Press.
Deleuze, G. (1994) *Difference and Repetition*. Translated by P. Patton. New York, NY: Columbia University Press.
Deleuze, G. and Guattari, F. (1987) *A Thousand Plateaus*. Translated by B. Massumi. Minneapolis, MN: University of Minnesota Press.
Deleuze, G. and Guattari, F. (1994) *What is Philosophy?* Translated by H. Tomlinson and G. Burchell. New York, NY: Columbia University Press.
Deleuze, G. and Parnet, C. (2007) *Dialogue*. Translated by H. Tomlinson and B. Habberjam. New York, NY: Columbia University Press.
Finley, S. (2018) Critical arts-based inquiry: Performance of resistance politics. In N.K. Denzin and Y.S. Lincoln (eds) *The Sage Handbook of Qualitative Research* (pp. 561–575). Los Angeles, CA: Sage.
Garoian, C.R. (2015) Performing the refrain of arts prosthetic pedagogy. *Qualitative Inquiry* 21, 487–493.

Gibson, C. (2018) Touring nation: The changing meanings of cultural tourism. In D. Rowe, G. Turner and E. Waterton (eds) *Making Culture: Commercialisation, Transnationalism, and the State of 'Nationing' in Contemporary Australia* (pp. 116–128). London: Routledge.

Giroux, H. (2013) The violence of organised forgetting: Thinking beyond America's disimagination machine. http://www.truthout.org/opinion/item.

Groys, B. (2008) *Art Power*. Cambridge, MA: MIT Press.

Hollinshead, K, (1999) Tourism as public culture: Horne's ideological commentary on the legerdemain of tourism. *International Journal of Tourism Research* 1, 267–292.

Hollinshead, K. and Suleman, R. (2018) The everyday installations of worldmaking: New vistas of understanding on the declarative reach of tourism. *Tourism Analysis* 23, 201–213.

Horne, D. (1992) *The Intelligent Tourist*. McMahon's Point, Sydney, Australia: Margaret Gee Publishing.

Jackson, A.C. and Mazzei, L.A. (2018) Thinking with theory: A new analytic for qualitative inquiry. In N.K. Denzin and Y.S. Lincoln (eds) *The Sage Handbook of Qualitative Research* (pp. 717–737). Los Angeles, CA: Sage.

Kirschenblatt-Gimblett, B. (1998) *Destination Culture*. Berkeley, CA: University of California Press.

Lanfant, M.-F. (1995) Introduction. In M.-F. Lanfant, J.B. Allcock and E.M. Bruner (eds) *International Tourism: Identity and Change* (pp. 1–23). London: Sage.

Leite, N. and Graburn, N. (2009) Anthropological interventions in tourism studies. In T. Jamal and M. Robinson (eds) *The Sage Handbook of Tourism Studies* (pp. 35–64). Los Angeles, CA: Sage.

Mavrič, M. and Urry, J. (2009) Tourism studies and the New Mobilities Paradigm (NMP). In T. Jamal and M. Robinson (eds) *The Sage Handbook of Tourism Studies* (pp. 645–657). Los Angeles, CA: Sage.

May, T. (2008) *Gilles Deleuze: An Introduction*. Cambridge: Cambridge University Press.

McKay, I. (2004) *Quest for the Folk*. Montreal: McGill & Queens University Press.

Mol, A. (2003) *The Body Multiple: Ontology in Medical Practice*. Durham, NC: Duke University Press.

Nye, J.S. (2004) *Soft Power: The Means to Success in World Politics*. New York, NY: Public Affairs.

Robinson, M. and Jamal, T.B. (2009) Conclusions: Tourism Studies – past omissions, emergent challenges. In T.B. Jamal and M. Robinson (eds) *The Sage Handbook of Tourism Studies* (pp. 693–702). Los Angeles, CA: Sage.

Rothman, H. (1998) *Devil's Bargains*. Lawrence, KS: University Press of Kansas.

Salazar, N.B. (2010) *Envisioning Eden: Mobilising Imaginaries in Tourism and Beyond*. New York, NY: Berghahn.

Saxena, G. (2015) Imagined relational capital: An analytical tool in considering small tourism firms' sociality. *Tourism Management* 49, 109–118.

Soanes, C. and Stevenson, A. (2006) *Oxford Dictionary of English* (2nd edn revised). Oxford: Oxford University Press.

Thomas, N. (1994) *Colonialism's Culture: Anthropology, Travel and Government*. Princeton, NJ: Princeton University Press.

Williams, J. (2009) *Gilles Deleuze's 'Logic of Sense': A Critical Introduction and Guide*. Edinburgh: Edinburgh University Press.

10 Afterword: Reflections on Paradoxes in Understanding, Culture, Mobility, and Tourism

Erdinç Çakmak, Keith Hollinshead and Hazel Tucker

'Let go of certainty. The opposite isn't uncertainty. It's openness, curiosity and a willingness to embrace a paradox, rather than choose sides'

Tony Schwartz

'How wonderful that we have met with a paradox. Now we have some hope of making progress'

Niels Bohr

The contributors to this book have sought to draw attention to some of the many paradoxes by, and through which, in tourism (and in Tourism Studies) peoples, places, pasts, and presents are variably understood and misunderstood. In particular, drawing attention to paradoxes and their prominence for tourism is of interest not only in Tourism Studies but also in other disciplines, including sociology, anthropology, cultural geography, as well as for many practitioner stakeholders, who design and consult in both the private and public sector. Indeed, tourism effects cultural change, and a myriad other forms of change, in ways which always, inevitably, will throw up new and inescapable paradoxes. Therefore, paradoxes are not to be avoided, for rather, we might welcome unfolding paradoxes for their help in better understanding the often contradictory and conflicting ways in which tourism 'does things' by producing ever-new dynamics, and ever-new complexities. As we noted in our introduction by referring to Kierkegaard: 'one must not think ill of the paradox, for the paradox is the passion of thought'.

Accordingly, our aims in this book were not only to exhibit some of the forms of paradoxical reasoning that may ordinarily flow through tourism and travel, but also to look at ways to think *with* the paradoxes, and to think *through* them. The function of the book is not to say what is right or wrong about a given scenario in tourism (or in Tourism Studies) but to enhance understanding of those who work in Tourism Studies/Tourism Management (and related fields). To this end, the implicit purpose of the book has not only been to ground a number of paradoxes in real-world situations, but to acknowledge that paradoxes themselves are 'dynamically capricious'. It is therefore useful to not only go beyond persistent, perhaps orthodox, notions of being vis-à-vis nations/peoples/traditions/spiritualities, but to expect that emergent notions of being are likely to be interpreted differently by different peoples at different times. The goal, then, in Tourism Studies/Tourism Management is to cultivate varying ways of looking at the culture of places and at the effects of globalisation and glocalisation on places.

As Gannon (2008: 9) has valuably expressed: 'thinking and reasoning about culture and globalisation [necessarily reflects the need for] a *nuanced* rather than a definitive approach. [When one thinks paradoxically] the world and its attendant issues become tones of gray rather than either black or white'. And so, heeding Gannon's judgement, in this book it has not been possible, nor was it the aim, to provide definitive solutions for real-world puzzles: rather, it was to hover over particular paradoxes in the hope that they can help explain why an encountered state of affairs exists and why a seemingly absurd or seemingly contradictory 'thing' continues. In this final chapter, then, we wish to underline some key 'paradox themes' in Tourism Studies that the book's chapters have highlighted. These themes relate to the paradoxes in understanding culture, mobility, and tourism representations/projections respectively, all central lines of paradoxical thinking throughout the book.

Understanding Culture: The Paradoxes of Here, There and Everywhere

The culture of a place, space, nation, or wherever, is an increasingly difficult thing to assess and define (Finley, 2018). While in many countries the ownership and control of cultural entities has become pointedly more centralised and convergent, policies of and about culture tend to involve more stakeholders than used to be the case. What is considered to be *culture* (or *cultural*) is assumed to be much broader these days than in recent decades and the term *culture* is itself no longer deemed to be something to do with the high arts and with forms of refined beauty and discerning knowledge. Rather, culture now is deemed to be something to do with an expanding range of *expressions* and *entertainments*. National or public policy on what relates to culture is not just what the particular government chooses to engage in but what it also chooses *not* to engage in vis-à-vis forms of 'critical citizenship' or 'nationing' (Rowe *et al.*, 2018; Turner, 1994). Cultural creations, museum displays, tourism presentations of heritage, are never neutral (Kirschenblatt-Gimblett, 1998). Public policy towards culture is therefore no easy thing to investigate, for policy processes tend to be intricate, unsystematic, and enigmatic here, there, and everywhere. While in some nations, central governments (such as those of China and France) like to engage in forms of direct control over culture, in other nations (such as the UK and the USA) the so-called management exercised by central government authorities tends to be an *at distance* affair delegated down to or out to quangos. Accordingly, it is difficult to obtain broad geographical consensus about what is defined as *culture* and/or *cultural* here, there, or anywhere, and very few ordinances or prescriptions about, for instance, *culture participation* or *culture development* readily cross unfiltered from one context to another.

In the light of this, it is the view of Bell and Oakley (2015: 64) that meanings of and about culture are always multiple and thereby that cultural policy is always inherently a polysemic affair. To Bell and Oakley there is never any watertight association between particular 'culture' and the particular 'nation' (or the particular 'place'). Yet, 'cultural forms', 'cultural activities', and 'cultural events' are ubiquitously harnessed in Tourism Management practice, and in Tourism Studies for that matter, to help define that nation, or place. Hence, under all these messy and mesmerising influences, in amongst tourism here, there and everywhere one may spot many paradoxes about culture. As Hobsbawm and Ranger (1983) have suggested, cultural traditions are not necessarily what has happened in the past over time and which thereby continue almost unblemished through to today, but they are (instead) what particular powerfully positioned groups *want to claim* 'has always happened longtime'. Paradoxically, then, traditions are not so much inherited cultural phenomena but are invoked politically charged phenomena. Furthermore, certain revered locales and cherished sites are normalised and naturalised to such a degree that esteemed parts of the nation (or the place) are turned into a kind of heritage museum via such dominant forms of pressure. And over time, paradoxically, where such political forces are strong, the resultant designated identity of a nation (or place) is not only something that emanates from the forms of culture exhibited there, but it can also *creatively produce* those very cultural forms afresh (Bell & Oakley, 2015: 119). Yet, there are few studies that consider the role of paradoxes in understanding the phenomenon of culture and what is inherited as cultural heritage by a community. The relevance and potentialities of turning paradoxes into a heuristic device for thinking through the relationship between tourism and *culture* and *cultural* have hopefully become evident through the chapters in this book.

Paradoxes in Understanding Mobility

The authors of Chapters 3 (Peyvel), 5 (Chen, Burns and J. Wang), 6 (Cheng), and 7 (de Vrieze-McBean) explore ways of deploying paradoxes in relation to the mobility of domestic tourists, foreign travellers, and exchange students. These chapters question 'traditional' modes of order, identity, power, and modernity in society and, by highlighting the paradoxes in the relevant case studies, adopt a more fluid perspective on cultural and personal subjectivity. This perspective is in line with Bauman (2013), who has produced inventive and provocative challenges to entrenched disciplinary ideas on *en groupe* identity and being, and which, in turn, have questioned orthodox ideas of order and stability in and across society. Concepts such as Bauman's *liquid modernity* have brought to the fore new modern and compulsive views on 'community', 'individuality', and emancipation. In addition, as highlighted in Chapter 9 (Hollinshead, Suleman, S. Wang, Nair and Vellah), Deleuze and Guattari (1987) see social and cultural life today as a multitudinous arena of new possibilities in which the global disorientations of our time have made available many fresh and previously unimagined opportunities for 'becoming' and for 'human flourishing'. To them, such are the new powers of possibility open to people today, including to social scientists and humanists who are charged with producing tenable ways of knowing about contemporary being.

To thinkers on global/glocal mobility, like Bauman and Deleuze/Guattari, it is crucial that the transformative effects of the new nomadisms of globalisation are closely monitored, for these modern impulses and imperatives have produced and are producing all manner of new affects/effects, new creative polysemic attachments, and new fluid aspirations. Throughout the chapters in this book the contributors highlight the paradoxes of contemporary mobilities in redefining 'identities' as new beings and becomings. The chapters thus emphasize the importance of examining the role of paradoxes in understanding the mobilities of all manner of 'travellers', not only for those who work in Tourism Studies but also for people working (and being and becoming) in all manner of tourism-related contexts, as will be expanded upon in the next section.

Paradoxes in Understanding Tourism: New Imaginative Projections of People and Place

The role of paradoxes in understanding tourism serves as a leitmotif throughout this book, and the authors of Chapters 2 (Eftychiou), 4 (Tan and Mura), 8 (Platenkamp) and 9 (Hollinshead, Suleman, S. Wang,

Nair and Vellah), in particular, explore emergent forms of subjectivity by way of matters of 'identity' outranked by matters of 'becoming'. They highlight some of the key paradoxes in understanding tourism and offer new imaginative projections of people and places which foreground the new fluidities of becoming different *with and among* new others, rather than being *fixed and contained*. Since tourism is a vanguard field and industry of globalisation (Fletcher, 2009), it is crucial that those who work in tourism studies stay abreast of such broader social science, humanities, and posthumanities insights on culture and identification under the global and glocal undulations of the 21st century. Thus, it is important to regularly gauge what the shortcomings of the imaginary of Tourism Studies might be, and what the particular *deficits of knowing* might be within both the tourism industry and the academy for the field. In other words, it is vital that those who work in the field of Tourism Studies/ Tourism Management appreciate the multiple and complex multilayered selves that tourism 'makes', thereby being open to, and open to cultivating, new sorts of aspiration and fresh realms of becoming.

The dynamic context in and around tourism and the fresh and fluid aspirational agitations of travel are likely to always be pregnant with paradoxical meaning and counteractive values. It is very hard to be still and rooted when one takes the self into those other scenes, those other sites, and those other storylines. Such are the dynamic and not readily predictable *fluctuations* of international travel, its related industries, and its effects. The politics and poetics of travel and tourism are no easy thing to fast corral and tightly conceptualise. Much *fluid acumen* (Hollinshead & Suleman, 2017) is thus called for, on the part of those who work in and study tourism, to understand these social, cultural, political, economic, environmental, and psychic undulations and projections.

Putting it Together: An Epilogue

The editors' hope is that that this book readily, perhaps even zealously, reflects the collage of contemporary paradoxes in tourism and tourism studies. In their various chapters the contributors have striven to question some of the contemporary paradoxes and contradictions co-deployed through the interactions between people and place for travel and tourism. The focus on paradoxes in the book was intended to stimulate oxygenated thought about travel/tourism and what it possibly and contrarily means, what it possibly and contrarily makes, and what it possibly and contrarily magnifies. The aim was not to be prescriptive and thereby batten down preferred understandings about the personal, societal, cultural power of tourism, but rather to undo the imaginative tethers and agitate richer reflective and reflexive thought about the contrarieties, the contradictions and the liquidities of our time.

Hopefully, this focus on paradoxes can further undiscipline the views which many of us have been taught to hold about what tourism is and can only be. Now, where shall I/we travel to next geographically … and why? Now, where shall I/we travel to next conceptually … and why? And why not?

References

Bauman, Z. (2013) *Liquid Modernity*. Cambridge: Polity.
Bell, D. and Oakley, K. (2015) *Cultural Policy*. London: Routledge.
Deleuze, G. and Guattari, F. (1987) *A Thousand Plateaus: Capitalism and Schizophrenia*. (Translated by B. Massumi). Minneapolis, MN: University of Minnesota Press.
Finley, S. (2018) Critical arts-based inquiry: Performances of resistance politics. In N.K. Denzin and Y.S. Lincoln (eds) *The Sage Handbook of Qualitative Research* (pp. 561–575). Los Angeles, CA: Sage.
Fletcher, J. (2009) Economics of international tourism. In T. Jamal and M. Robinson (eds) *The Sage Handbook of Tourism Studies* (pp. 166–187). Los Angeles, CA: Sage.
Gannon, M.J. (2008) *Paradoxes of Culture and Globalization*. Los Angeles, CA: Sage.
Hobsbawm, E. and Ranger, T. (1983) *The Invention of Tradition*. Cambridge: Cambridge University Press.
Hollinshead, K. and Suleman, R. (2017) Politics and tourism. In L.L. Lowry (ed.) *The Sage Encyclopedia of Tourism* (pp. 961–964). Los Angeles, CA: Sage.
Kirschenblatt-Gimblett, B. (1998) *Destination Culture*. Berkeley, CA: University of California Press.
Rowe, D., Turner, G. and Waterton, E. (2018) *Making Culture: Commercialisation, Transnationalism, and the State of Nationing in Contemporary Australia*. London: Routledge.
Turner, G. (1994) *Making it National: Nationalism and Australian Popular Culture*. St Leonards, NSW, Australia: Allen & Unwin.

Index

Note: References in *italics* are to figures, those in **bold** to tables.

academia 128, 132, 133, 141
Achilles and the Tortoise 129
affect 151, 153
agents of modernity 19
Akama, J. 6
Amoamo, M. 135
Anglo-Western centrism *v.* non-Western imperatives 4–5
Ap, J. 94
Appiah, K.A. 134, 135
Archer, M.S. 111
Arendt, H. 142
Argyrou, V. 20, 23–4, 29
Aristotle 128–9
Arlt, W.G. 120, 121
art 147, 148, 151, 156
Ashcroft, B. *et al.* 6
Asian tourism 3
aspiration 7, 150
assemblage 154
Augé, M. 153
Australia 154
authenticity xv, 1

Baffie, J. 45
banalization 45
Bartilet, J.L. 111
Bateson, G. 130
Bauman, Z. 8, 153, 162
Beck, U. 23
becoming 162
Bell, D. 116, 161
Bellah, R.N. 118
bidirectionality 152
Bogue, R. 151
Bohr, N. 159
Boland, R. 142
Bourdieu, P. 25, 30
Braidotti, R. 154–5
Breda University of Applied Sciences, Netherlands 139–40
Bruner, E. 152
Buck, E. 145

Burawoy, M. *et al.* 15
Burma 140
business models 77
Butler, G. *et al.* 54

Çakmak, E. 7, 23
capitalism 115–18, 149–50, 155
Carroll, L. 129
CDA *see* critical discourse analysis
center-periphery models 35
certainty 159
Chen, H. *et al.* 90
Chen, Y. *et al.* 76
China: Ministry of Culture and Tourism (MCT) 114, 115, 116
China National Tourism Administration (CNTA) 75, 114–15
Chinese Outbound Tourism Reseach Institute (COTRI) 120
Chinese outbound tourism to Europe 10–11, 114–15
  as product of capitalism 115–18
  as product of Communism 116, 120–2
  as product of Confucianism 118–19
  as product of consumerism 123–5
  soft power 115
  conclusion 125–6
Chinese racial discourse 103
  contemporary racial discourse 108–9
  discourse of superior Han Chinese 104–6
  historical hierarchical race relations 106–8, *107*, *108*
  racial discourse defined 103–4
  reflexivity 110–11
  discussion 109–11
  see also cross-cultural encounters
Chinese students in London *see* Chinese racial discourse; cross-cultural encounters
Chinese Tourism Law (CTL2013) 10, 74–6
  consumer mentality 90–3
  group package tours (GPTs) 76

Index

interviewee profiles **80–1**
low-fare package tours (LFPTs) 74–5, 76–8, 77
'Measures for the Administration of Tour Guides' (2018) 94
methodology 78–82
moral self-regulation 93–4
paradoxes influencing effectiveness 86–94
perceived (in)effectiveness of 82–6
principles-based regulation approach 86–90
conclusion 94–5
Chinese tourists
international expenditure of 3, 4
in Thailand 2
see also Chinese outbound tourism to Europe; Chinese racial discourse
Chouliaraki, L. 58
Clifford, J. 133, 153
CNTA *see* China National Tourism Administration
Cohen, E. 1, 5
Cohen, S.A. 5
Colebrook, C. 146, 148, 149, 150, 152
colonialism 6, 23, 40, 45, 60, 64
colonization *v.* post-colonization and decolonization 5–6
colony, the 40
communicative paradoxes 86
Communism 114, 116, 120–2
Communist Party of China (CPC) 116–17
compliance paradoxes 89
Confucianism 38, 106, 114, 116, 118–19
consumer mentality 90
consumerism 34, 45–6, 114, 123–5, 149
contradictions 3, 115, 154
COTRI (Chinese Outbound Tourism Reseach Institute) 120
counter-narrative 6
COVID-19 pandemic 3
creative production 161
critical discourse analysis (CDA) 55, 57–9, 58, 65–6, 66
cross-cultural encounters 10, 97–8
educational tour production 98–100
positionality 98
racism 99–101
students' experiences 100–3
conclusion 111–12
see also Chinese racial discourse
CTL2013 *see* Chinese Tourism Law
CTO *see* Cyprus Tourism Organisation
CTrips 119

cultura franca 142
cultural identity 154
cultural relationships 4
culture 160–2
Cyprus 9, 15
colonial Cyprus 18–19
encountering paradoxes 15–18
Kakopetria 24–8, 25, 27
mass tourism 19–22
modernity 19–22, 23–6, 28–9
paradoxes 23–9
postcolonial Cyprus 19–22
'reflexive tourism' 22–3
Tourism Development Ofiice 19
tradition 17, 23, 24–9
Troodos region 16, 18–19, 21–2, 23, 24–5
women 17, 24, 27–8
conclusions 29–30
Cyprus Tourism Organisation (CTO) 17, 23

Dann, G.M.S. xv, 138
darshana 152
Darwin, C. 128, 129, 133
decolonization 5, 6, 132
deficits of knowing 162
Deleuze, Gilles 11, 146, 162
and the dogmatic image of thought 147–52
hidden and unexpected truths 145–6
and the 'opening up' of Tourism Studies 152–5
and Tourism Studies refreshed 146–7
summary: call for dynamic genesis 155–7
Deng Xiaoping 116, 121
Denton, K.A. 121
Desforges, L. 54
destination culture 153
devil's bargains (of tourism) 153
DeVrieze-McBean, E.R. 119, 142
diagonal impacts 152
Dikötter, F. 106
discourse 19, 21–22, 59, 103–6, 108–9
disruption 151
distanciation 137
Doane, A. 104
dogmatic image of thought 147, 150
'dynamic genesis' 145–6

Edensor, T. 6
empowerment 6–7
enforcement paradoxes 89
Engels, F. 120, 122

England 154
English as lingua franca (ELF) 130–2, 142
English language 59–60, 67
Escobar, A. 21
Euromonitor International 123, 124
exoticism 43

Fairclough, N. 58, 65
fantasmatics 135
Fauna 109
Feng Shui 37
Finley, S. 156
Fishwick, C. 103–4
flourishing 156, 162
fluid acumen 162
Foucault, M. 6, 30, 110–11, 130
framing 53
Frazier, R.T. 109
Frissen, P. 129
Fulcher, J. 117–18

Gadamer, H. 136
Gannon, M.J. 160
Garcia, J.L.A. 104
Garoian, C.R. 155
gaze, the 26, 30, 53
geophilosophy 11
Germany 120, 122
Gibson, C. 154
Giddens, A. et al. 23, 132
Giroux, H. 156
global ethnography 15
globalization 2–3, 7, 8–9, 11, 35–6, 45, 132, 152, 160, 163
globally standardized vs. locally dynamicized 8–9
Goldman Sachs Global Investment Research 124
Gonzales, M. 65
Goodpaster, K.E. 78
Graburn, N. 34
Gramsci, A. 6–7
Groys, B. 156
Guattari, F. 147, 148, 149, 151, 162
Guba, E.G. 134

Halbwachs, M. 67
Hall, C.M. 35, 53
Hall, S. 104
harmony 125
Hayashi, Fumiko 46
Hébrard, Ernest 41
hegemony 7, 15, 21, 53
Henderson, J.C. 53
hereness 145

heritage 51, 61, 64–7
heritage tourism 6
Hevi, E.J. 106
Hobsbawm, E. 161
Hollinshead, K. 6, 122, 162
Horne, D. 145, 146, 152–3
Huayuan International Travel 119
Hughes, M. et al. 3–4

identity 6–7, 21, 56–7, 63
ideological difference 67
ideological legerdemain 145
ideologies 21
imaginary, the 36
imaginative projections 162
immanence 148
impacts of policy 62
impacts of tourism 151–2
incongruence 4
intensity 149
International Council on Monuments and Sites (ICOMOS) 51
International Sociological Association xiii
international tourism academia 11, 128
    academic development in a network-society 132–3
    academic insulae and the route to Ithaca 133–4
    English as a lingua franca 130–2, 142
    knowledge production 136–7
    Kumasi on Ithaca 134–7, 139, 140, 141
    paradoxes and Russell's types 128–30
    three modes of knowledge production 137–41, *138*
    conclusion 141–3
interpretation 139
interpretive paradoxes 86
Isaac, R.K. et al. 140, 141

Jaakson, R. 6
Jamal, T.B. 146
Jenness, D. 18
Jia, Y.Q. 77
Jiang Zemin 116
Jing, Lou 108–9
Jóannesson, G. et al. 4
Jobanputra, D. 117

Kammas, M. 18
Key 109
Kierkegaard, S. 1, 160
Kirschenblatt-Gimblett, B. 145, 153
knowledge production 136–8

Korean Airlines 142
Kotler, P. et al. 114
Kwek, A. 118

Larsen, S. xv
Lee, Young-Sook 124
Lenin, V. 120
Levuka, Fiji 136–7
Lin, R. 105
liquid modernity 162
Liu, P.H. 106
localism 8–9
Lost in Thailand (movie) 2
Lowenthal, D. 67

madeness 145, 153
McCabe, S. 110
MacCannell, D. xv, 1
McGehee, N.G. 93
McKay, I. 145
McKinsey Global Institute 123
Magazine Littéraire 130
Malaysia Book of Records 70
Malaysia's Catholic mission schools 10, 50–2
    'accepted' identities and heritage 53–4
    adopting values, adapting cultures 60
    architecture, mission school heritage and tourism 51, 68–71, 69
    Cameron Highlands Convent 69
    critical discourse analysis (CDA) 55, 57–9, 58, 65–6, 66
    English language 59–60, 67
    George Town 68
    government policy 62–3
    Guild of Assisted Catholic Schools 50, 63, 67
    'heritage' status 51, 52, 68
    interview focus 55–7
    Malacca 51, 52, 68
    mission school heritage 61–2
    narratives and participants 54–5, 55, 59
    new identities 63–4
    'potential energy' 64–5
    preservation via tourism-linked change 66–8
    St Joseph's Novitiate, Penang 70, 70
    St Paul's Church, Malacca 51, 52
    symbolism and form 60–1
    discussion 65–8
    conclusion 71–2
Mao Zedong 106, 107, *107*, 120, 121
Marcus, G. 133
Markides, K.C. et al. 24
Marschall, S. 6

Marx, K. 120, 122
Mavrič, M. 153–4
May, T. 146, 149, 150, 156
MCT *see* China: Ministry of Culture and Tourism
Meyer, M. 57
Milillo, N. 117
Minca, C. 4
mobility 8, 153, 155, 160, 162
modernity 19, 24
MOFAA 106, 107
moral self-regulation 93
Morgan, N. 5
multidirectionality 152
multiple modernity 9
multiplicity 152

narratives 54, 59, 132
National Geographic (magazine) 45
nationing 161
'nationing' of cultural identity 154
neo-colonialism 53
New Mobilities Paradigm 153
Nisbett, R. 132
nomadic theory 154
Nye, J. 115

Oakes, T. 4
Oakley, K. 161
object multiple, the 157
Observer (newspaper) 142
OECD (Organisation for Economic Cooperation and Development) 7, 141
ontology 4, 6, 116, 125, 146
Other, the 29
overtourism 3

Palmer, C. 54
palpation 150
paradigms 135, 137
paradoxes 129–30
    defined xv, 146
    freedom 130
    reflections on 11–12, 159–60
    Russell's types 129–30
    tourism paradoxes xv–xvi, 1–4
    understanding culture 160–2
    understanding mobility 162
    understanding tourism 162–3
    epilogue 163–4
Parks, T. 131, 132
Parnet, C. 157
passion of thought 1, 160
People's Daily Online 120

Persianis, K.P. 18
Philippou, N. 18
philosophy 11, 128–9, 130, 143, 148, 150
Platenkamp, V. 133, 140, 141
political apparatus *v.* peoples'
    empowerment 6–7
polysemy 162
Popper, K. 129, 130
Portegies, A. *et al.* 139–40
positionality 98
postcolonialism/postcoloniality 21, 35,
    42, 53
posthumanism/posthumanity 156
post-colonization 5–6
post-disciplinary research 140
potentia 147
power relations 6–7
pradakshina 152
preservation 51, 67
Pretes, M. 65
Prideaux, B. *et al.* 77–8
primitivism 42
Pritchard, A. 5
progress 159
Publius Terentius Afro 135

Rabinow, P. 30
racism 99–101
    see also Chinese racial discourse
Ranger, T. 161
RC50 (Research Committee 50)
    International Sociological
    Association xiii
re-making (via tourism) 122
real/reality 34, 53, 153
Red Tourism 120–2
reflections on paradoxes 11–12, 159–60
reflexive discourse 21
reflexivity 110
regulation approaches 87
representation 6, 132, 147
Republic of Cyprus 19–20, 21
rhizomatics 151–2
Robinson, M. 146
Robson, I. 78
Robson, J. 78
romanticism 45
Rothman, H. 145, 153
Rowe, D. *et al.* 161
Rushdie, S. 135
Russell, B. 128, 129–30, 136, 137, 143

Sachdeva, S. *et al.* 93
Said, E.W. 30, 134
Salazar, N.B. 152

Saxena, G. 154
Schwartz, T. 159
Seaton, T. 138
Shalvi, S. *et al.* 93
Singapore Art Museum 70
singular identities *v.* liquid/plural
    aspirations 7–8
Soanes, C. 2, 4
soft power 156
soft science 135
Southgate, B. 67
Speitkamp, W. 122
Spinoza, B. 148
staged authenticity xv
stakeholders 78
Steiner, G. 131
Stevenson, A. 2, 4
structure of the book 9–12
Suleman, R. 145, 162
Sunstein, C.R. 75, 88
sustainability 16–17, 23
symbolism 60

Thailand
    Chiang Mai 1–2
    low-fare package tours 77
thinking/thought 29, 146, 150, 156, 160
Thomas, N. 145
Tito, Josip Broz 120
Tourdust 117
tourism paradoxes xv–xvi, 1–4, 162–3
tourism power 156
Towson, J. 124
Tribe, J. 133
trust paradoxes 90
truth 134
Tselichtchev, I. 116
Tsien, Roger Yonchien 105
Tucker, H. 5, 6, 35, 53
Turner, G. 161

UNESCO World Heritage sites 33–4, 51,
    52, 68
United Nations
    Sustainable Development Goals
        (SDGs) 97, 111
    World Tourism Organisation 2015
        (WTO) 3, 97, 111
universality 5
University of Cambridge 15, 105
Urry, J. 26, 153–4

value judgment 56, 60
Vecco, M. 65
Veijola, S. *et al.* 4

Vietnam 9–10, 33–5, 40
    Bà Nà 39, 40, 45, 46, *46*
    Buddhism 37
    colonial period 35, 36, 40
    consumerism 34, 45–6
    Dà Lat 36, 39, 40–1, 42, 45, 46
    domestic tourism 33
    French romanticism revisited 45–6
    gardens and prayers 36, 37, 38–40, *39*
    hill stations 35, 36, 39, 40–2
    Lào Cai 43, 44
    methodology 36
    mountains for the Kinh 34, 35, 36–7, 39–40, 42
    Perfume Pagoda 37, *38*
    postcolonial geography 35–6
    primitivism and exoticism 42–5, *44*
    Sa Pa 36, 39, 40, 41, 42, 44, 45
    Tam Dao 36, 37, 39, 40, 41, 42
    UNESCO world heritage sites 33–4
    women 43, 44, 45
    conclusion 46–7
virtual possibility 147
virtual reality 149

Wallerstein, I.M. 8, 29
Wang Yong 115

webs of encounter 155
Weick, K. 110
Welz, G. 23
West, C. 93
western hegemony 15, 20–1, 29–30
Williams, J. 146
Winter, T. 3
Woetzel, J. 124
women
    Cyprus 17, 24, 27–8
    Vietnam 43, 44, 45
Wong, K.K.F. 94
worldmaking 135, 145

Xi Jinping 121
Xu, Y. 93

youth tourists 1, 2

Zeno of Elea 129
Zhang, J. 5
Zhang, L. 109
Zhang, Q.H. *et al.* 78
Zhong, C.-B. 93
Zhou Enlai 106–7
Zhou, Q. 120